LONDON'S UNDERGROUND SINCE 1985

Front cover image

On 8 July 1995, three trains of 1959 Stock are present at Finchley Central Station, at which Northern Line trains from the south are terminating owing to engineering works to the north of it on the High Barnet branch. The train on the far right is working a shuttle service to and from Mill Hill East.

Back cover images

Oddities on the Central Line for many years were the units of Cravens-built 1960 Stock, consisting of two such driving motor cars with older refurbished trailers between them. Most of their working lives were spent on the Woodford to Hainault and Epping to Ongar shuttles. On 2 November 1989, one of them has arrived at Woodford Station.

Still looking very smart, despite being due for replacement by new S7 Stock trains, a train of District Line D Stock calls at West Ham Station on 29 June 2014. Driving motor car 7129 brings up the rear.

LONDON'S UNDERGROUND SINCE 1985

A Journey in Colour

JIM BLAKE

PEN & SWORD TRANSPORT

AN IMPRINT OF PEN & SWORD BOOKS LTD.
YORKSHIRE – PHILADELPHIA

First published in Great Britain in 2024 by
Pen and Sword Transport
An imprint of
Pen & Sword Books Ltd.
Yorkshire - Philadelphia

ISBN 978 1 39905 558 1

A CIP catalogue record for this book is available from the British Library.

Typeset by SJmagic DESIGN SERVICES, India.

Printed and bound in India by Parksons Graphics Pvt. Ltd.

Pen & Sword Books Ltd incorporates the imprints of Pen & Sword Books Archaeology, Atlas, Aviation, Battleground, Discovery, Family History, History, Maritime, Military, Naval, Politics, Railways, Select, Transport, True Crime, Fiction, Frontline Books, Leo Cooper, Praetorian Press, Seaforth Publishing, Wharncliffe and White Owl.

For a complete list of Pen & Sword titles please contact

PEN & SWORD BOOKS LIMITED
George House, Units 12 & 13, Beevor Street, Off Pontefract Road,
Barnsley, South Yorkshire, S71 1HN, England
E-mail: enquiries@pen-and-sword.co.uk
Website: www.pen-and-sword.co.uk

or

PEN AND SWORD BOOKS
1950 Lawrence Rd, Havertown, PA 19083, USA
E-mail: uspen-and-sword@casematepublishers.com
Website: www.penandswordbooks.com

INTRODUCTION

THE PERIOD THAT this second photographic survey of the London Underground covers, again using photographs from my archives, began with the system at what perhaps was its lowest ebb. The Conservative government had snatched control of it away from the Greater London Council at the end of June 1984, and under the overall control of the government 'quango' London Regional Transport, a new entity dubbed London Underground Limited was created. Its remit was to reduce operating costs and, before too long, such in-house facilities as catering and cleaning were tendered out to private companies, with of course the lowest bidders winning and often leading to shoddy performances, well below the time-honoured level of service that had been provided under unified London Transport command. Cuts amongst operating staff were made too, with an all-out drive towards one-person operation (OPO), and thus the abolition of guards.

The net result of all this led to a visible decline in standards on the Underground. Trains began to gain an uncared-for appearance, particularly the older unpainted ones, and this was made much worse by graffiti, which began to spread like the plague at the same time. Stations became targets of graffiti vandals too, whilst litter became a problem at many of them.

A particular memory I have from three years or so after this decline set in is of walking through one of the Underground's lesser-used subways connecting a sub-surface booking hall with one of central London's main line termini and seeing workmen filling a disused lift, whose gates were open, with wastepaper – both general litter and obsolete posters stripped from the subway walls. So shocked was I at the sight, and also intrigued by seeing the old lift doors open, that one of the workmen said, 'Oi, what you looking at, mate?' I declined to answer, and when a few days later the news of the awful King's Cross fire disaster hit the headlines, I was doubly horrified; the station in whose subways all that inflammable waste paper was apparently stored was King's Cross! This disaster must have been exacerbated by cuts in cleaning costs, if not actually caused by them, and when eventually the enquiry results were published, heads rolled at London Underground headquarters. That fateful date, 18 November 1987, was truly one of the worst days in the entire history of the system. Subsequently, various measures were taken to ensure that such a tragedy did not happen again; just too late, sadly, for the thirty-odd poor souls who perished there, as well as the many more badly injured.

Meanwhile, operating staff morale was at rock-bottom, as I know first-hand from several friends working on the system at the time. Line after line was converted to OPO, until the last guards of all were withdrawn when its last victim, the Northern Line, succumbed in January 2000. However, by that time there was, perhaps literally, 'light at the end of the tunnel'!

After Margaret Thatcher was 'deposed' by her own party late in 1990, hers was replaced by a more benign Tory regime. Then, in 1997, the Labour party was re-elected to government after 'eighteen years of Tory misrule'. Legislation was swiftly enacted to re-establish a London-wide local authority, amidst whose remit came once again overseeing local transport within its boundaries, both on the buses and the Underground. The new Greater London Authority renamed this entity Transport *for* London, and although much time and public money was consumed by a very cumbersome part-privatisation of Underground maintenance operation imposed by central government in its early years, a better system has since emerged.

Meanwhile, during the 1990s the Jubilee Line was finally extended eastwards, though not on the alignment originally intended. Brand new rolling stock, the 1996 Stock, was introduced to it, sending the unsuccessful 1983 Stock to an early grave, whilst similar OPO 1995 Stock replaced the Northern Line's complement of 1956, 1959/62 and 1972 Mk1 Stock in the last years of the twentieth century, completing the job a few weeks into the twenty-first. The Central Line had, meanwhile, had all of its 1962 Stock replaced in the first half of the 1990s by new OPO 1992 Stock, examples of this also working the Waterloo & City Line which London Underground inherited from British Rail, along with the trains themselves, at the same period.

Minor losses from the Underground system took place in 1994, with the closure of the little-used Piccadilly Line Holborn to Aldwych branch, and the Central Line's Epping to Ongar protrusion into rural Essex. These have been compensated for by not only the extension of the Jubilee Line mentioned above, but also a short extension of the Northern Line from its Charing Cross branch's Kennington loop to Battersea Power Station, opened in September 2021. At the time of writing, other proposed extensions are frustrated, as always, by lack of funding.

All of the photographs included are my own, presented in chronological order; very many of them having never been published before. During my years as a prolific transport photographer (1961-2015), I usually concentrated on rolling stock, or stations and services, due for withdrawal or closure, and this will be evident in many of those presented here. I have not dwelt on the technical aspects of the rolling stock illustrated, as that is not this book's purpose, such details being readily available elsewhere. For readers interested in such things, I would recommend the excellent publications produced by the London Underground Railway Society.

May I thank Colin Clarke for scanning all of my 100,000 or so negatives a few years ago, thereby making it easy for me to locate the pictures included herein, and also John Scott-Morgan of Pen & Sword Books for arranging this book's publication.

JIM BLAKE
Palmers Green
July 2024

GENERAL OVERVIEW OF PASSENGER ROLLING STOCK TYPE ALLOCATION AS AT 1 JANUARY 1985

A. SUB-SURFACE LINES

Circle Line - C69/C77 Stock
District Line - D78 Stock plus C69/C77 Stock on Edgware Road-Wimbledon service
Metropolitan Line (main section) - A60 & A62 Stock
Metropolitan Line (East London Line) - A60 & A62 Stock
Metropolitan Line (Hammersmith & City Line) - C69/C77 Stock

B. TUBE LINES

Bakerloo Line - 1938 Stock, 1959 Stock, 1972 Mk2 Stock
Central Line - 1962 Stock (main service); 1960 Stock (Woodford to Hainault shuttle).
Jubilee Line - 1972 Mk2 Stock, 1983 Stock
Northern Line - 1956 Prototype & 1959 Stock, plus odd 1962 Stock units
Piccadilly Line - 1973 Stock
Victoria Line - 1967 Stock

With the impending final demise of the last 1938 Stock units in service fast approaching, a farewell tour was organised using the 'Starlight Express' unit on 19 May 1985. This view finds it heading south through the northern car sheds at Queens Park Station headed by driving motor car 11012. Somewhat oddly, Bakerloo Line trains in normal service also pass through them on this track.

After switching over at Baker Street to the Bakerloo Line's former tracks now occupied by the Jubilee Line, the tour proceeded up to Stanmore, from where it followed the Metropolitan Line to the northern side of the Circle Line. From there it continued via the Hammersmith & City Line to Whitechapel and Barking, continuing further to the East along the District Line to its easternmost extremity of Upminster Depot which, of course, is some distance east of the station. Here, driving motor car 10291, whose unit has also been adorned with cream side window pillars, leads the formation, providing an interesting contrast with one of the District's recently-delivered D78 Stock trains.

***Above*: Later in** the day, the tour finds itself again contrasting with a D78 unit, when passing westwards through Gloucester Road Station, with driving motor car 11012 at its trailing end. This section of the tour took it around the southern side of the Circle Line. Of note here are the former District Railway destination indictors, still in use.

***Opposite above*: The last** train of 1938 Stock on the Bakerloo Line, and as then intended on the entire Underground system, ran on 20 November 1985. A few weeks beforehand, on 11 October, one of the last remaining trains in service approaches Kensal Green Station, bound for Stonebridge Park. Unfortunately, I no longer have record of which car is leading this train. Even on their final day, rumours were circulating that they would *not* yet disappear from the Underground, which proved to be correct!

***Opposite below*: Notwithstanding the** rumours, a final, final 1938 Stock Farewell Tour was operated on 30 March 1986, more than four months after the 'last' such train had run in public service. I rode on this one, which began in unfamiliar territory – at the Central Line's White City Station, where driving motor car 11012 is at its trailing end as it collects its passengers. By this time, it was common knowledge that the trains would indeed re-enter service, but on the Northern Line.

235

D
SPECIAL
1938 STOCK
UNDERGROUND
FAREWELL TOUR
11/84

***Above*: The train** headed westwards along the Central Line towards West Ruislip, but instead of continuing to the terminus, ran around the west side of the Central Line Ruislip Depot, pausing to allow those on board to photograph some of the interesting rolling stock always stabled there. Along with some typical flat wagons used on engineers' trains, three typical London Transport battery locomotives are visible in this view on what was a dismal, rainy day. On the left is L59, in the centre L16 and on the right L47. Although appearing identical, they have different pedigrees – L59 is a Pickering-built example dating from 1952, L16 was built by Metro-Cammell in 1970, whilst L47 dates from 1974 and was built by British Rail!

***Opposite above*: After passing** Ruislip Depot, where several 'withdrawn' 1938 Stock units were visible awaiting repaint and reconditioning for the Northern Line, our tour took us onto the Metropolitan Line's Uxbridge branch, with which a connecting track had recently been provided from the Central Line. However, we did not initially go to Uxbridge, but instead gained the Jubilee Line track into Baker Street, in order to take us on to the Bakerloo Line. Here we have arrived at its Northern terminus, Harrow & Wealdstone, where the train heads for the reversing siding with driving motor 10291 in the lead, before returning south. I am standing in the exact spot where the two locomotives (Jubilee 4-6-0 45637 *Windward Islands* piloting newly rebuilt Princess 4-6-2 46202 *Princess Anne*) heading the Down express landed after striking the wreckage of the initial collision of the disaster on 8 October 1952, demolishing the station footbridge and falling across the southbound Watford DC and Bakerloo Line. A unit of 1938 Stock was approaching when this happened, cutting off its current and preventing, perhaps, it striking the wrecked locomotives. Had the tube train been there just a few seconds sooner, they may well have landed on top of it, thus compounding an already horrific disaster.

***Below*: Having returned** to Baker Street and regained Jubilee Line metals, the tour then took us to Stanmore, back to Wembley Park, and then around the Metropolitan Line on which we did reach Uxbridge this time. Driving motor car 11012 heads the train in its centre platform, and will lead us along the Piccadilly Line, where of course such units ran in service for many years, as far as Acton Town.

***Above*: From Acton** Town, we reversed once more and gained the Piccadilly Line's Heathrow branch. An unexpected highlight was that the train took us around the newly-built loop to Heathrow Terminal 4 Station, where we were allowed to alight and photograph our train. The station did not open to the public until 12 April, so we were the first fare-paying passengers to sample it. After this, the tour returned to Acton Town, finally ending at Ealing Broadway District Line terminus. A thoroughly enjoyable day despite the horrible weather.

***Opposite above*: Two eras** of Underground history are evident in this picture on 11 May 1986. On the left are the splendid new platform buildings of Dollis Hill Station, dating from the Bakerloo Line's 1939 takeover of the Metropolitan Stanmore branch, and stopping services at such intermediate stations between Finchley Road and Wembley Park. On the right, A62 Stock driving motor car 5217 speeds by with a typical Metropolitan Line formation of two A60/A62 units bound for Baker Street. These trains represent the line's modernisation in the early 1960s, A60 Stock having replaced loco-hauled and T Stock compartment trains on its main line, and A62 Stock completing the job by replacing F Stock and P Stock on the Uxbridge and Watford branches. Now, the line was soon to convert to OPO, and its trains severely afflicted by graffiti.

***Opposite below*: By 18 August** 1986, OPO has befallen the Metropolitan line, the most obvious visual sign of this being the alteration to one of the A Stocks' cab windows as evident in this view. Also, major engineering work involving track renewal on the 'fast' tracks shared with the then threatened British Rail services between Marylebone and Aylesbury is being carried out, as this scene near Pinner shows.

Above**: The train** in whose cab I was riding when taking the previous picture, thanks to a friend who was its motorman, was headed by A60 Stock driving motor car 5092, which has arrived at its Watford terminus. My friend, now long retired, may be seen changing ends in the distance. In common with many drivers when lines converted to OPO, he hated not having a guard to chat to at such times, and shortly afterwards transferred to the Northern Line, where he had the added incentive of the chance to drive 1938 Stock before too long.

Opposite above**: Watford Met.** terminus has an island platform, with one siding either side. On the same occasion, A60 driving motor car 5026 heads an eight-car formation stabled in one of them. The station is somewhat remote from Watford town centre, and in fact the Metropolitan Railway did acquire land for a terminus in the High Street but was unable to secure the route to it owing to Cassiobury Park being in the way. The building meant to be its terminus passed into LPTB ownership, being sold in the mid-1930s. It became J.D. Wetherspoons' Moon Under Water pub in the 1990s, and its frontage bears a resemblance to various Metropolitan Railway stations built, or rebuilt, in the 1920s. Meanwhile, despite very advanced plans having been made, also in the 1990s, to divert the branch to connect with the disused former British Rail Croxley Green branch and terminate Metropolitan Line trains at Watford Junction, they have foundered, largely due to its being outside the Greater London Authority's area.

Opposite below**: Meanwhile, I** enjoyed another cab ride from Watford all the way to Aldgate, and am now heading west again as far as King's Cross. A60 Stock driving motor car 5086 heads a train bound for that point approaching Farringdon Station. In the distance on the left, a British Rail class 317 EMU is working the Bedford to Moorgate service on the former Metropolitan Widened Lines, which was largely superseded by the new Thameslink services in June 1988.

WATFORD

ALDGATE

***Above*: On the** same evening as the previous few pictures, a three-car Piccadilly Line 1973 Stock unit headed by driving motor car 864 has arrived at the Aldwych terminus of the short branch from Holborn, for many years reduced to single track. The second disused tunnel has been used for storage purposes, with its Aldwych platform often used to try out new station designs over the years. Detailed plans were made in the mid-1960s to extend the branch beneath the Thames to Waterloo, but though 'dusted off' subsequently, these came to nought. The branch finally closed at the end of September 1994, the last nail in its coffin being the expense of renewing its antiquated lifts, unjustified for such a little-used station.

***Opposite above*: On 22 September** 1986, driving motor car 1292 is at the trailing end of a Bakerloo Line service arriving at Willesden Junction Station. This part of the complicated station is referred to as the 'new' station, having been built by the London & North Western Railway to separate the Bakerloo and Watford DC electric service from their main line. It originally comprised two outer platforms for through trains, plus two terminal platforms for trains terminating from the south. Platforms on the main line part of the station, where semi-fast services running from Euston (and Broad Street in rush hours) to and from stations north of Watford used to call, were removed when the main line was electrified in the mid-1960s.

***Opposite below*: The next** Bakerloo Line train southbound is a 1972 Mk2 Stock. Originally new to the Northern Line, these were transferred to the Bakerloo following that line's receipt of 1959 Stock. However, when the new Jubilee Line took over the Bakerloo's Stanmore branch in May 1979, they were resettled there. When, in turn, the Jubilee Line received 1983 Stock, they moved back to the Bakerloo, eventually replacing all 1959 Stock which returned to the Northern Line. At the time I took this picture, no one could have foreseen that these trains would still be working the Bakerloo Line at the time of writing (winter 2022/23), and more than likely for at least another ten years, too! Even now, they are the oldest non-heritage railway rolling stock in regular service in Britain!

Also on 22 September 1986, two trains of Central Line 1962 Stock meet at North Acton Station in the evening rush hour. Nearest the camera is driving motor car 1619, heading an Epping-bound service. 1959 and subsequently 1962 Stock served the Central Line for some 34 years, being finally ousted by 1992 Stock in 1995. The station is very much in original condition, having been built originally by the Great Western Railway in 1923.

Whereas the Central Line's 1935-1940 New Works Programme extensions, interrupted by the Second World War, to West Ruislip and to Epping and Ongar respectively, were all completed (with the exception of a short stretch beyond West Ruislip to Denham), those of the Northern Line similarly interrupted were not. It defies logic that the extension of the Northern City Line from Drayton Park and Finsbury Park to Highgate and Alexandra Palace and to join the main Northern Line at East Finchley – through areas fully built up in Victorian and Edwardian times – was abandoned when so much work on it had been completed. This view, on 5 October 1986, shows the substation beside the line at Crouch Hill which, contrary to what has been published elsewhere, was built and fully equipped for the extension. Of note at the bottom of the wall facing the camera are the holes where the cables emerged, now filled in, since the building after many years' disuse has been adopted by Islington Council's Crouch Hill Community Centre. All necessary cabling, conductor rails and signalling equipment to complete the line's electrification had been delivered to site before further work was placed in abeyance in 1941. Understandably, much equipment was removed for reuse elsewhere to repair bomb damage during the war, though some conductor rails remained in place until the extension was officially abandoned in the early 1950s, as did all of the equipment in this substation, and the one at Muswell Hill Station. This was removed and later reused for such far flung projects as electrifying the Central Line between Epping and Ongar, and the Metropolitan Line between Rickmansworth and Amersham. This of course made a complete mockery of the 'Green Belt' excuse used to abandon the extension in North London! The substation at Muswell Hill was demolished in the early 1960s, but this one at Crouch Hill had an extra storey built atop it a few years ago.

***Above*: Another relic** of the uncompleted Northern Line extensions seen on the same day is the signalbox built to supervise the lines converging at Park Junction, where the Alexandra Palace branch parted company with the 'main' Northern Heights branch to Edgware and High Barnet, as well as Highgate Depot and Highgate Wood sidings situated between them. Following the abandonment of the extensions and the complete removal of the branch to Alexandra Palace in 1956/57, followed by the destruction of the branch southwards to Finsbury Park in 1971/72, all it had to deal with was the depot itself and the approach to East Finchley Station. With the depot being closed and 'mothballed' in March 1984, the whole site soon became illicitly accessible from Highgate Wood, with the singalbox falling prey to vagrants, drug addicts and worse. This view is taken from its control room level, looking towards the 'main' branch whose tracks are visible on the left. It was not needed when the depot reopened in 1989, and finally demolished in the mid-1990s, never having been fully used for its intended purpose.

***Opposite above*: Equally depressing** is this view of the western end of Highgate West Tunnel, taken on the same occasion. It clearly shows how the exit from the northbound bore had been bombed during the war, and rebuilt with a circular portal, rather than the original elliptical shape. A London Transport publication praising their staff's herculean efforts during the war to keep services going featured this rebuild, emphasising that trains were running again through the tunnel within a couple of days after a quick repair. However, London Transport tube trains *never*, of course, ran in service through this tunnel, only a much-reduced LNER service shuttling between Finsbury Park and Alexandra Palace. Of course, much goods traffic used the line in those days too.

***Below*: On one** of the rare occasions when London Underground's battery locomotives are seen in daylight at a station still in passenger use, L33, a solitary London Transport-built example dating from 1962 and renumbered from L76 in 1974, is stabled in the centre platform at Golders Green on Saturday, 11 October 1986. As may be observed in the background, extensive track renewal work is being carried on south of the station that weekend, and trains from the north are reversing there, using the southbound track only, which still gives them access to the adjacent depot. A train of 1959 Stock is just visible heading for it on the left.

***Above*: 1972 Mk1 Stock driving** motor car 3239 arrives in the usually southbound platform to take up service, whilst a crane is in action lifting heavy equipment as works progress. No doubt the battery locomotive had brought it to the site.

***Opposite above*: By now,** some units of 1962 Stock, originally delivered for Central Line services, have found their way onto the Northern Line. Also on 11 October 1986, driving motor car 1744 leads the single four-car unit, calling at Hendon Central on the shuttle service provided between Edgware and Golders Green. Normally, these worked in multiple, being compatible with 1956 and 1959 Stock of course.

***Opposite below*: Also at** Hendon Central that day, 1959 driving motor car 1057 trails in a single four-car set heading north. A comparison with the previous picture clearly shows the difference in position of the red stop light on each stock. Perhaps the dates for the heavy engineering work at this time were chosen to coincide with the Jewish high holy days of Rosh Hashanah and Yom Kippur at this time, when traffic was lighter on the Edgware to Golders Green section.

GOLDERS GRN

***Above*: Services are** running normally on the Northern Line's High Barnet and Mill Hill East branches as driving motor car 1276 brings a train of 1959 Stock bound for Morden via Charing Cross into East Finchley Station on sunny 6 November 1986. Visible to left of centre in the background is the former Great Northern signalbox which once controlled movements to and from the station goods yard on the right, now long since converted to a car park. British Railways goods yards on this branch closed early in 1964, finally having been served in their last few years by diesel locomotives via the branch from Finsbury Park which was part of the uncompleted Northern Line extensions.

***Opposite above*: Ten days** later, on a somewhat grey Sunday, 16 November 1986, the open sidings on the west side of the Northern Line's Morden Depot appear at first sight to be dominated by 1959 Stock. However, close scrutiny reveals that the second train from the right is headed by a 1962 Stock driving motor car; the different stop light position gives it away!

***Opposite below*: Sadly, the** curse of graffiti was beginning to plague the Underground by this time. The 1956, 1959 and 1962 Stock on the tube lines, and the A Stock on the Metropolitan Line in particular, already looking somewhat careworn, were prime targets this vandalism. On 20 November 1986, tags have defaced this Northern Line unit at Morden Station – sadly, efforts to cleanse it have not been very successful.

MORDEN
← WAY OUT

***Above*: A happier** spectacle on the same occasion is presented by a train headed by one of the units of 1956 Prototype Stock entering Morden Depot. These can immediately be distinguished by the old-style destination indicator lights on the cab fronts. Both on the Piccadilly and the Northern Line, these worked in multiple with 1959 Stock units.

***Opposite above*: Even smarter,** the first train of 1938 Stock to reappear on the Northern Line is also at Morden that day. In this view, driving motor car 11012 faces the camera. Their return to service, a full year after 'final withdrawal' from the Bakerloo Line, merited a substantial slot on LWT's *Six O'Clock Show*. Several enthusiasts, including myself, were featured and my old friend David Rowe had pride of place; he had recently transferred to the Northern Line especially to drive 1938 Stock trains, and was filmed with this unit in Morden Depot. Their return to the Northern Line was necessitated by an increase in services, which had perhaps been cut just too drastically beforehand.

***Opposite below*: Ten days** later, on 30 November, the same 1938 Stock train with driving motor car 10291 at its other end heads for Kennington via Charing Cross in the centre platform at Golders Green, already having lost the shine given to it for its TV appearance. This and its partner 11291, both originally LNER property, became unit 007 on the Isle of Wight, and upon their final withdrawal in January 2021, became part of the Isle of Wight Steam Railway's museum collection, restored in LT livery.

MORDEN

A
KENNINGTON
CHARING X
NORTHERN LINE
024

***Above*: Over at** East Finchley on the same day, 1956 Prototype Stock driving motor car 1008 is at the trailing end of a service bound for Morden via Bank. These units were withdrawn during the 1990s, effectively replaced by 1962 Stock transferred from the Central Line.

***Opposite above*: A interesting** scene beneath the bridge carrying Pursley Road over the trackbed of the abandoned Mill Hill East to Edgware branch on the same day finds its path being cleared of foliage. In the foreground are cable runs, fitted but never used, for the Northern Line's extension over the whole of this branch, whilst it may be clearly seen how the bridge had both been rebuilt to accommodate two tracks, as well as for the road it carries to be widened. Foliage is being cleared to allow the creation of a footpath alongside Copthall Park, in much the same way as the Parkland Walk occupies the other part of the uncompleted Northern Line extensions between Finsbury Park, Highgate and Muswell Hill.

***Opposite below*: Not** only did staffing and cleaning suffer under the LRT 'quango' but so, it seems did signage on the Underground. At Highbury & Islington Station, whoever produced this sign stuck on the booking office window has made a faux pas: it is, of course the Great Northern service it should be referring to, not the Great Eastern! The date is December 1986; British Rail Eastern Region services had taken over the ill-fated Northern City Line just over ten years previously.

ON SUNDAYS THE
GREAT EASTERN
ELECTRIC SERVICE
RUNS FROM KING'S CROSS
OR FINSBURY PARK STATIONS

***Above*: By 28 December** 1986, another complete train of 1938 Stock had been spruced up and returned to service on the Northern Line. It too has been given cream window pillars in the same style as the 'Starlight Express' unit. Headed by driving motor car 10205, it noses out of Morden Depot's car sheds, flanked by two units of 1972 Mk1 Stock. Along with partner 11205, this car would become Isle of Wight unit 004, and following their withdrawal on the island early in January 2021, is being converted to a cafe (restored to LT livery) at Holliers Park, Branstone.

***Opposite above*: Although showing** a destination plate for High Barnet, driving motor car 11012 and the rest of a seven-car train of 1938 Stock is actually reposing in Golders Green Depot on 18 January 1987.

***Opposite below*: On the** same bitterly cold day, when we had had a fair amount of snowfall, driving motor car 1749 of one of the 1962 Stock units transferred to the Northern Line from the Central Line contrasts with units of 1972 Mk1 Stock in the sidings at Golders Green Depot. It is a Sunday, so plenty of units are 'at home'. Perhaps this was the usual time that graffiti vandals attacked the train as here on 1749's cab front. Thankfully, much stricter security prevails at Underground depots nowadays than it did then.

D
HIGH BARNET
NORTHERN LINE
BANK
381

SPECIAL
000
749

Also in Golders Green's snowy sidings that Sunday, 1972 Mk1 Stock driving motor car 3513 has also been attacked by taggers, whilst the car next to it resembles the reverse of a blackboard in an infants' class!

Another view of the car in the previous picture reveals it to be 1959 Stock driving motor car 1031, detached from its unit and also missing at least one of its passenger windows. It had been withdrawn following accident damage and was being cannibalised for spares.

More doleful sights are to be seen at Ealing Common Depot's 'Alps' sidings on 21 March 1987. Standing beside some thoroughly vandalised CO/CP cars dumped there for several years is a 1938 Stock driving motor car converted for depot shunting as a pilot motor car, which is rapidly suffering the same fate. Regrettably, I am no longer sure which one it is – perhaps L142.

Somewhat surprisingly also looking very sorry for itself at the depot that day is TCC1, one of the two 1938 Stock driving motor cars converted to run with the tunnel cleaning train, looking very different from its immaculate self when displayed at the Acton Works LT Golden Jubilee open day less than four years earlier. It was, however, subsequently 'resurrected'.

***Above*: Sadly, Central** Line trains are also prey to graffiti as the cab of driving motor car 1635 shows when heading a train of 1962 Stock at White City on 18 March 1987. Of note here are the unusual station name signs, as visible on the right. White City was a new station opened in 1947, replacing the awkward and antiquated Wood Lane station nearby.

***Opposite above*: Another** seven-car train of 1938 Stock has re-entered service on the Northern Line by 25 March 1987, with driving motor car 10255, one of the formerly LNER-owned ones, bringing up the rear on a service from Edgware at Brent Cross Station. My two little daughters Margaret and Felicity have enjoyed their ride on it! This car, and its partner 11255, would become unit 008 on the Isle of Wight, and remain in service until December 2020. They are being restored by the London Transport Traction Group at present, at the unlikely location of the Llanelli & Mynydd Mawr Railway.

***Opposite below*: Caught by** my camera through the trees of Cherry Tree Wood, just south of East Finchley Station on 9 May 1987, one of London's Transport's Plasser & Theurer tamping units, dating from 1980, is attending to the track on the former GNR/LNER branch leading to Highgate and Finsbury Park. The depot at the former point had been closed and 'mothballed' nearly three years earlier but would not reopen until January 1989.

A
EDGWARE
CHARING X
006

***Above*: Out in** the East, on 12 June 1987, 1960 Stock driving motor car 3908 is at the trailing end of a Woodford to Hainault shuttle service, about to depart from the former station for a run around the northern side of the Hainault loop. By now, the two Standard Stock trailers in their unit formations have been replaced by a single 1938 Stock car.

***Opposite above*: With the** need to replace the ageing 1956/59/62 Stock trains within a few years becoming apparent, London Underground had three new four-car tube-size units, all to the same general design, delivered in 1986/87 for test purposes with this in mind. Dubbed 1986 Stock, they were each given distinctive liveries, with a view to bringing 'coloured' trains back to the system, perhaps as a means of deterring graffiti vandals. Two were built by Metro-Cammell, being the red and green-liveried units, whilst the third, which was blue, was built by British Rail at Derby. All three were displayed in Woodford Station's third, terminal, platform on 12 June 1987 to assess public reaction to them. This is the 'red train', as it was known.

***Opposite below*: The three** units, which only had a driving cab at one end, each had different electrical equipment although were to the same basic design. Different passenger seating styles were tried out too, which spectators were asked to sample. The trailer cars also had windows in each end, evidently as a safety consideration at a time when the plague of mugging and even sexual assault was also afflicting the Underground, a situation not helped by the spread of OPO. This view shows the green train and the blue train facing each other when on display. As will be evident by their appearance, the design of the Central Line's 1992 Stock was derived from these units. They saw passenger use on the Jubilee Line in 1988/89, in two six-car formations. After that, and storage for several years, all were scrapped except for the driving motor car of the green train which was saved by the LT Museum.

SCOTCH
BLOCK

FIRST TRAIN

***Above*: A unit** of 1972 Mk1 Stock was also present at Woodford that day, perhaps for comparative purposes with the new experimental cars, or maybe it had merely brought them there. At any event, such trains were never to operate on the Central Line. In this view of its driving motor car 3503 facing south, a 1960 Stock and a 1962 driving motor car are just visible in the distance, representing 'the old order'.

***Opposite above*: In sad** and dire contrast to the new cars being displayed on the adjacent platform, Cravens 1960 Stock driving motor car 3902 has been marred by graffiti. It has just discharged its passengers from the Hainault to Woodford shuttle, and heads for the sidings south of the station.

***Opposite below*: This close-up** view of 4929, one the 1938 Stock trailers used to replace Standard Stock working with the Cravens 1960 Stock driving motor cars on the Woodford to Hainault shuttle, shows how although it has been painted 'silver', it does not quite match the two unpainted aluminium 1960 Stock cars. Wear and tear on all three cars is evident, but at least the red paint on the trailer is not showing through – yet.

WOODFORD
2
161

4929
D

Above**: With driving** motor car 1538 nearest the camera, heading a West Ruislip-bound service, an interesting comparison is provided at Woodford Station between two of the Central Line's 'run of the mill' 1962 Stock trains and the experimental units on display on the right.

Opposite above**: This view** showing the approaches to the Northern Line's Edgware terminus on 5 July 1987 captures a train of 1959 Stock arriving in its platform two. The apparent wasteland in the foreground is part of the intended trackbed for the extension to Bushey Heath, which in its final plans would have crossed from the branch from Finchley Central and Mill Hill on a new overbridge (which was actually installed) above the branch from Golders Green, in order to avoid conflicting track movements. In the distance, another train of 1959 or perhaps 1962 stock is standing on one of the sidings which occupied part of the alignment towards Mill Hill, after the extensions were abandoned.

Opposite below**: This view** looking north from the same spot shows, hiding in plain sight, the additional island platform built at Edgware for trains to and from the Bushey Heath extension. A builders' hut stands at its southern end, and cable racking has been placed in the southbound trackbed. The signal cabin on the right was also built in connection with the extension, and therefore never fully utilised, just like the one at Park Junction, Highgate illustrated earlier, whose design it resembles. In order to accommodate this new construction, the LPTB's Edgware bus garage had to be demolished and rebuilt further away from the railway premises – yet more wasted work on the uncompleted Northern Line extensions!

***Above*: Inside Edgware** Station's original covered part, but on 14 July 1987, a Northern Line train of 1959/62 Stock has been marked with graffiti, but in order to maintain the service, has had to be put in use. Non-driving motor car 9745 of the 1962 Stock unit of the train awaits departure from Platform Two.

***Opposite above*: Whereas most** graffiti besmirching our Underground trains, stations and infrastructure is unintelligible gibberish, that defacing the 'amidships' of driving motor car 1278 of the same train's trailing unit leaves little room for doubt, though it is not represented in this image. A number of public complaints were made about it, especially as strict instructions existed stipulating that trains must not leave their depots in such a state. Meanwhile, this view shows that the roof over the two platforms does not extend all the way to the entrance/exit area.

***Opposite below*: 1962 Stock driving** motor car 1744 heads it out of Edgware Station bound for Morden via Bank. This view also shows how that station still has an unfinished appearance, some forty-six years or so after the link from Mill Hill East and extension to Bushey Heath was abandoned. In the meantime, the platform on the left (now named Platform One) had been brought into regular use. Presumably, the covered roof should have extended over this and also the two Bushey Heath platforms too.

15
101

FUCKLONDON TRANSPORT

At Edgware that afternoon, a train of revived 1938 Stock arrived, and I had the pleasure of travelling on it all the way to Morden. With driving motor car 11116 awaiting a return journey to Edgware, it stands at that southern terminus. Even this car has graffiti, above its cab windows. Interestingly, it had had its driving cab completely demolished, as well as severe damage otherwise, in a collision at the Bakerloo Line's Queens Park Depot in 1949. Presumably, since additional cars of 1938-style stock were to be produced at that time, that is how it came to be rebuilt. Along with its partner 10116, it also saw further use as unit 003 on the Isle of Wight, though was an 'early' withdrawal in 2000.

***Above*: One of** three former City & South London Railway (C&SLR) stations still with their original narrow island platforms at this time, Clapham North Station has one of its platform name signs removed, showing its original name, Clapham *Road*. It had been renamed in 1926, when the line was extended to Morden. Have vandals removed the plate showing its current name?

***Right*: An oddity** at Finchley Central Station, pictured on 16 July 1987, is this large sign in Station Road, on the northbound side of the station itself, directing passengers to it. It is mounted on one of London Transport's standard bus stop posts, and the message in its panel explains that it is *not* a bus stop and shows where local buses actually do stop. The structure behind it is a former Great Northern Railway ticket office for passengers wanting to travel to stations to either Edgware or High Barnet, long since in use as a greengrocers' shop.

***Above*: So the** old song goes, 'Finchley Central is two-and-sixpence from Golders Green on the Northern Line...change at Camden Town', but I could not do that on this occasion, since both of its stations at Euston had been put out of action by a serious flood a couple of weeks beforehand. So I had to use the No.13 bus, which was more direct anyway! Trains from the Edgware branch ventured only as far as Hampstead, to where 1959 Stock driving motor car 1139 leads this train in Golders Green's centre platform. It wrongly shows a blind displaying 'Hampstead via Charing Cross', meant of course for trains terminating there from the south. The blotches on the cab front show where LT staff have made valiant efforts remove graffiti.

***Opposite above*: Correctly showing** plain 'Hampstead', fellow 1959 Stock driving motor car 1103 heads another service bound for that station, in the platform at Golders Green alongside the depot. It is unusual for trains to terminate at Hampstead from the north.

***Opposite below*: Trains on** the High Barnet and Mill Hill East branches terminated at East Finchley, whose two centre platforms for once saw quite a lot of passenger use. Here, 1959 Stock driving motor car 1167 is at the trailing end of a train which has terminated in what should have been the northbound platform for trains from Moorgate to High Barnet via Finsbury Park. Replacement bus services had to be arranged to cover both of the Northern Line's inner sections.

HAMPSTEAD
006

EAST FINCHLEY
WAY OUT
034

By 2 August 1987, services are back to normal on the Northern Line, and driving motor car 10255 heads a smartly turned-out train of 1938 Stock calling at East Finchley. It has come from Morden via Bank on a through journey to Mill Hill East. What a shame it cannot continue to Edgware, and perhaps Bushey Heath, as had been intended!

On the same occasion, a Sunday as the almost empty station car park suggests, a unit of 1956 prototype stock with driving motor car 1011 leading heads a train bound for Morden via Charing Cross. Its destination blind is difficult to read, perhaps the glass, the blind itself – or both – need cleaning.

A splendid relic of the Underground's earlier days is Kilburn Park Station on the Bakerloo Line, pictured on 2 October 1987. Opened in 1915, it is in a similar style to the famous Leslie Green stations on this as well as the Northern and Piccadilly Lines but was designed by Stanley Heaps and does not have the upper floors and larger semi-circular windows typical of the former. Sadly, fly-posting has defaced the corner of the building on the left.

In stark contrast, an ugly concrete bridge, built for a one-way system at the junction of Hanger Lane, the Western Avenue and the North Circular Road, frames this view of 1962 Stock driving motor car 1408 heading a Loughton-bound Central Line train into Hanger Lane Station on 7 November 1987. On the right are the tracks of the former GWR/BR Western Region main line to Birmingham and Birkenhead, totally defunct today at this point.

***Above*: Back on** the Northern Line, the partially covered platforms at its Edgware terminus are occupied by two extremes for the line at the time – a train of 1938 Stock, and a train of 1972 Mk1 Stock. The date is 12 April 1988. Apart from some cars of the latter stock used to bolster the Victoria Line's 1967 Stock fleet, which lasted until 2012, and others used for similar purposes with the Bakerloo Line's 1972 Mk2 trains which are still in service today, all 1972 Mk1 trains had been withdrawn by the end of 1999, *before* the last 1959/62 Stock units had gone.

***Opposite above*: At the** trailing (northern) end of the 1938 Stock train is driving motor car 11012, this interior view of which shows the guard's door control panels. His bag is placed by the communicating door to the next car. Soon we will be off.

***Opposite below*: At the** first stop southwards, Burnt Oak, the guard looks out from his train to make sure everyone has safely alighted and boarded. He helpfully held the doors for a few seconds to allow me to hop off and take this photograph.

***Above*: As luck** would have it, another train of 1938 Stock was in action at the northern end of the Northern Line that day, and on the Edgware branch too. It is led by driving motor car 11184, which unusually is lacking a destination plate – presumably it is bound for Kennington via Charing Cross as it approaches Brent Cross Station. Along with partner 10184, this became unit 001 on the Isle of Wight just over a year later, though was withdrawn in 1999 and scrapped in 2006.

***Opposite above*: The other** unit comprising the train pictured above had driving motor car 10229 at its trailing end, which unlike the leading unit was refurbished at Ruislip Depot. Along with partner 11229, it also went to the Isle of Wight but not until 1992, becoming unit 009. It remained in service until 2016, thereafter being used for shunting at Ryde, St. John's Road Depot. This car survives at the West Somerset Railway. In the aftermath of the tragic King's Cross fire of November 1987, the Northern Line's 1938 Stock trains, which included a fair amount of woodwork considered a fire risk, were withdrawn in May 1988. I thought as a result that my ride on this unit a month or so previously was my last on one – but family holidays on the Isle of Wight in the 1990s proved otherwise!

***Opposite below*: Sadly, the** sudden withdrawal of the refurbished 1938 Stock which had returned to the Northern Line only some eighteen months previously meant that they never returned to Highgate Depot in normal service. This view of that establishment on 30 October 1988 shows, in the foreground, track lifted from the redundant Highgate Wood sidings on the right. However, work is progressing to bring the depot shed back into use, and it reopened in January 1989.

EDGWARE
VIA CHARING X
NORTHERN LINE
004
A

***Above*: The Northern** Line's Camden Town Station boasts one of the most complicated interchanges and junctions on the entire Underground system, all on the same line too. Here, trains from the separate branches running beneath Central London, via Bank and Charing Cross respectively, link and then trains from both diverge to continue to the Edgware and High Barnet/ Mill Hill East branches respectively – and vice-versa. The actual track junctions are south of the station, which itself has an intriguing network of passages enabling interchange between each of the northern branches. The station is undergoing refurbishment on 15 April 1989 as 1959 Stock driving motor car 1063 leads a High Barnet train that has come via Charing Cross.

***Opposite above*: In June** 1989, the centenary of the Metropolitan Line's Chesham branch was celebrated by running a special train hauled by ex-BR Western Region 9400 class 0-6-0PT 9466, with appropriate headboards, to and from Chalfont & Latimer. Here, elaborate measures involving a water tanker and a long hose, ensure the loco's tanks are well-filled at Chesham terminus.

***Opposite below*: The same** train, formed using a Southern Region 4VEP EMU in Network SouthEast livery, had the veteran Metropolitan electric locomotive No.12 *Sarah Siddons*, at its London end, in order to avoid the loco changing ends at termini. It departs from Chesham, where the bay platform dating from its pre-1961 steam days is now trackless. Perhaps a pannier tank which had actually been used on the Metropolitan Line, albeit only on goods trains, might have been a better choice of engine; after all more than one of these is preserved. However, the event was a resounding success, and paved the way for the regular 'Steam on the Met' events a few years later.

SPL
CHESHAM
1889
YEARS
1989
100
9466

10

***Above*: On 20 September** 1989, a train of Hammersmith & City Line C69 and C77 Stock approaches Paddington, the leading unit (of four C69 cars) completely dwarfs one of the single-car diesel units that replaced steam on Western Region local services in the late 1950s, which is working a through service to and from the Greenford loop. A class 47 (Brush 4) diesel hauling an express into Paddington completes the picture.

***Opposite above*: The little** single unit has taken me to Greenford, where a central bay platform is the terminus of this service, between the two Central Line platforms. Nowadays, the Greenford loop service only ventures as far as West Ealing, and at the time of writing former District Line D Stock units are proposed to operate it. Meanwhile, a train of Central Line 1962 calls on its way to West Ruislip.

***Opposite below*: On a** misty 2 November 1989, on the easternmost reaches of the Central Line, the two Craven-built 1960 Stock driving motor cars adopted for LT's track testing train some twenty years earlier revisit their old haunts when standing in Woodford Station, now renumbered L132 and 133 in the non-passenger fleet. The latter is nearest the camera here, having originally been car 3905. L132 was 3901.The unit's two Standard Stock trailers are long gone, however, replaced by a trailer of 1973 Stock thrown spare from the Piccadilly Line following fire damage to the rest of its unit. This is numbered TRC666, and is the actual track recording car.

2
L 133

***Above*: On the** same occasion, the two Cravens 1960 Stock cars of the track testing train are reunited with their fellows on the Woodford to Hainault shuttle. Their smart red, white and black livery contrasts with the dirty grey, graffitied state of those in public service, as this view graphically shows. Fortunately, this situation was to be remedied before too long on this stock, as will be illustrated later.

***Opposite above*: Also on** a drizzly and grey 2 November 1989, 1962 driving motor car 1580 has arrived at the Central Line's Ongar terminus. Of note is how its car number has been repositioned, and the line name added. Gone is the traditional London Transport name amidships on each driving motor car, replaced by a London Underground roundel. London Underground have recently re-introduced a daily, all-day, service to the Epping to Ongar branch in attempt to increase its patronage, the little-used Blake Hall Station having closed eight years previously. This train will work a 'short' to North Weald, where the removal of the second track used for such workings now made no difference, since only one unit was in use on the shuttle. Sadly, it was all to no avail, and the service was reduced again in the spring of 1990.

***Opposite below*: At the** other extremity of the Central Line, West Ruislip Station, fellow 1962 Stock driving motor car 1575, has arrived with a train which will return to distant Epping on 5 April 1990. Here, it retains its London Transport fleetname, and its fleet number in the traditional place for this stock but has gained a Central Line name.

NORTH WEALD
1580
Central
Line
171

EPPING
Central
Line
1575
065

***Above*: Captured through** the window of one of the cars in the train pictured above, two withdrawn Standard Stock pilot motor cars stabled outside Ruislip Depot have suffered the full force of graffiti vandals. This has completely obscured their stock numbers, as well as covering their windows. Unfortunately, stock stabled for any length of time in the same position inevitably falls prey to this problem. I have managed to identify the worst stricken car as L131, formerly 3693, converted to a pilot motor car in 1967.

***Opposite above*: Illustrating how** the newly-built (or rebuilt) stations on the Central Line's western extension to West Ruislip completed after the war were produced on a far less grand scale compared with those on the Piccadilly Line and Northern Line extensions of the 1930s, another train of 1962 Stock calls at Northolt Station on 7 April 1990. This was to reduce costs, and use of scarce manpower and materials, in the period of austerity in the early postwar years. What a shame about the £3 million or so (at 1939 values) wasted on the uncompleted Northern Line extensions!

***Opposite below*: Perhaps in** an effort to defeat the graffiti vandals attacking the ageing A60/62 Stock trains, this four-car unit of A60 Stock has been given a new livery of red cab front, grey roof with blue window surrounds and white skirt. Despite the name 'Metropolitan Line' being placed above the passenger windows behind the cab side doors, it is working the East London Line when awaiting departure from New Cross Gate on 20 April 1990. The Met. still supplied A60/A62 rolling stock for the line until the end of its London Underground existence in December 2007.

WEST RUISLIP

WHITECHAPEL
1173

Despite being smartened up at Ruislip Depot (as the 'R' logo shows below its stop light), 1962 driving motor car 1619 still bears the scars of graffiti as it emerges into daylight at the eastern end of Stratford Station in May 1990 heading for West Ruislip. By now the Docklands Light Railway is in operation, and various improvements have been made to Stratford Station, notably the installation of lifts as shown on the left.

On the same day, a gala day was held at Epping Station, to mark the 125th anniversary of its opening in 1865. To mark the occasion. this unit of 1960 Stock, working the Epping to Ongar shuttle, has been smartly adorned in traditional Underground livery. What a difference from their appearance illustrated earlier!

In the same formation, 1938 Stock trailer 4927 has regained its original livery, and very smart it looks too. The train appears to be full, perhaps a 'first' in the history of this section of the Central Line since its electrification in 1957! In this smart livery, the remaining 1960 Stock trains ended their career working the Epping to Ongar shuttle when it was finally closed at the end of September 1994.

More typical of the Central Line, of course, are the trains of 1962 Stock which continue to suffer from graffiti. On the gala day, when a bus rally and transport enthusiasts' bazaar (which members of the North London Transport Society had a hand in organising) were held in the station car park, driving motor car 1701 awaits departure from Epping to West Ruislip. Within just over five years, sights such as this would be a memory.

***Above*: Summer sunshine** in July 1990 shows to good effect the graceful Charles Holden design of East Finchley Station in this view looking south. But why are there so many people on both of its generally disused centre platforms, and why are there sheets bearing current and 1930s Underground symbols across the tracks? Has the 'missing link' between here and Finsbury Park finally been brought into use?

***Opposite above*: Wrong! (Of course).** The event was to unveil new liveries for the Northern Line's rolling stock, as a prelude to commemoration of the centenary of the C&SLR, the first deep-level tube railway, and progenitor of the Northern Line. Complete seven-car trains of 1959 and 1972 Mk1 Stock were painted in an approximation of 1930s Underground livery, and a new style of red cab front and passenger doors, white body and roof and blue skirt, respectively. Driving motor cars 1044 and 3203 face the camera, having been driven side by side from Highgate Depot for the occasion. The adornment of the 1959 Stock train would remain unique, but the style of the 1972 Mk1 Stock would not only be used when refurbishing all Mk2 Stock of this type, but also for 1967 and 1973 Stock tube units, as well as the sub-surface A, C and D Stock units.

***Opposite below*: Midnight down** the 'Dilly, August 1990: towards the end of traffic hours, engineers' trains may be observed passing through deep-level Underground stations on their way to work sites to be occupied during the night. They are usually 'topped and tailed' by battery locomotives, of which L27 (one of the batch of eleven built by Metro-Cammell for London Transport in 1965) is at the trailing end of this train, passing through Piccadilly Circus Station when I was awaiting my northbound Piccadilly Line train homewards from a good drink in The Kings Arms, Poland Street, Soho.

SPECIAL
SPECIAL

BAKERLOO LINE
PLATFORMS 1 & 2 STRAIGHT ON
BAKERLOO LINE
PLATFORMS 1&2
27

***Above*: 1990 also** marked the 90th anniversary of the opening of the Central London Railway, from which today's Central Line grew. On the last weekend of October, its vast Ruislip Depot was thrown open to the public. For some years, overhauls and major repairs of rolling stock, principally the Central Line's own, had been devolved there from Acton Works. Here, several 1962 Stock cars are undergoing attention with driving motor car 1479 nearest the camera.

***Opposite above*: An oddity** on display at Ruislip Depot is L84, a Unimog road rail vehicle, whose wheels can be raised or lowered when necessary to allow both modes of travel. It is one of two used for shunting and conveyance or stores in and around the depot. Three of these were obtained during the 1980s, however such hybrid vehicles were not a new idea, the LMS, for example, having a road-rail compatible bus for use on branch lines more than fifty years previously.

***Opposite below*: Representing a** new breed of battery locomotive, L67 and L63 (just visible on the left) are also on display at Ruislip Depot. Numerically the last of five supplied by Metro-Cammell in 1985/86, they were evidently not as successful as their predecessors of basically 1930s design, many of which remain in use today.

46
45
44
LONDON

44
43
40
39
38
67

Back inside Ruislip Depot, L151 is one of two 1938 Stock driving motor cars converted in 1977 as a weed-killing unit, having originally been 11327; its partner 10327 working in tandem with it, renumbered L150 to spray tracks infested with weeds.

As mentioned above, the basic body design of London Transport's battery locomotives dates from the 1930s, and also in Ruislip Depot, L18 is one of three perpetuating it supplied by Metro-Cammell in 1971. It is still in use at the time of writing. On the adjacent track is an immaculately turned out 1938 Stock car that has been recently overhauled.

The immaculately turned out car in the above picture is ballast motor car L147, converted in 1976 from 1938 Stock driving motor car 11034, which is paired with L146, formerly 10034 with which it had also been paired when in service. One of the cranes used by London Underground for heavy engineering work stands next to it.

In total contrast is another 1938 Stock ballast motor car, L144 converted from 10257. It still bears the earlier maroon livery used for non passenger-carrying stock, and by the coat of dust upon it, appears to have been out of use for some while!

***Above*: Perhaps the** oldest item of self-powered rolling stock in Ruislip Depot's extensive sidings that day is electric sleet locomotive ESL117, one of those converted from two former Central London Railway driving motor cars joined back-to-back after removal of their passenger sections. It dates from 1903 and was converted in 1940.

***Opposite above*: On show** at the Ruislip Depot open day too is the British Rail-built 1986 Prototype Stock 'blue train'. In the event, it and its fellows never ran in service again, and it seems odd in retrospect that they could not have found use on the Waterloo & City Line, though of course London Underground did not take this over until 1994 when it was, perhaps, just too late.

***Opposite below*: Representing the** current generation of London Underground road-bound emergency vehicles, Mercedes-Benz 814 demountable truck 2491M is at Ruislip depot. It is carrying the emergency permanent way track team cabin. Interestingly, its fleet number continues in the series established by London Transport for non-passenger road service vehicles before the war – and this series is still in use today.

Emergency
H223 FLB

***Above*: Some interesting** moves took place in the autumn of 1990, with the 'repatriation' of Standard Stock tube cars to London Underground from the Isle of Wight. They had been sent there during the winter of 1966/67 to replace steam traction on the island's surviving line between Ryde Pier Head and Shanklin, the last having been withdrawn in the spring of 1990 – replaced by 1938 Stock. Some cars had already been taken to Ruislip Depot in time for the open day at the end of October 1990, and now in November, battery locomotive L56, one of three produced for LT by Pickering in 1951, brings a train of five more of them home.

***Opposite above*: The 'cortege',** which did indeed travel at almost walking pace, was transferred from British Rail to London Underground metals at Wimbledon, from where it was towed along the District Line to Earls Court. These pictures show it passing through Putney Bridge Station. As may be observed, two of the cars were repainted externally in 1920's 'Morden Edgware Line' (the Northern Line's name upon the extension of former C&SLR and Hampstead tube complex after combination) livery, the repainted car nearest the camera being one of two trailers in the formation.

***Opposite below*: The other** repainted car is the only surviving control trailer of the type, and also the last to be used as such, both on the Isle of Wight, and in London Transport service, when replaced by 1938 Stock on the Northern City Line in October 1966. Both this car and the trailer in NSE livery visible on the left, latterly numbered 5279 and 7296 respectively by LT, await full restoration at LT Museum's Acton Depot. The other cars in the formation are two driving motor cars, both in Network SouthEast livery. Sadly, both of these, along with the other trailer, were scrapped in 2013 after being held at Acton Works for possible preservation.

MORDEN-EDGWARE LINE

1789
49

***Above*: Trailing the** 'cortege' through Putney Bridge Station is battery locomotive L58, another Pickering-built example and one of three new in 1952. When the train reaches Earls Court, it will reverse and be headed by L58, following the District Line to Ealing Common Station and then the Piccadilly Line to Ruislip Station, from where it will take the connection into Ruislip Depot.

***Opposite above*: Some unusual** sights were to be seen in the Piccadilly Line's Northfields Depot at this time, too. Nearest the camera is 1973 Stock uncoupling non-driving motor car 390, one of two cars which are pilot refurbishments for the type. Of particular note are the windows at the car ends, meant to allow what happens within them to be visible from adjoining cars, as a deterrent to criminals attacking lone passengers late at night. The fleet was not, however, refurbished until 1996-2001, and also done by outside contractor Bombardier, rather than in-house.

***Opposite below*: Also in** Northfields Depot is 1973 Stock driving motor car 889, which had been seriously damaged in a fire in the tunnel between Bounds Green and Wood Green Stations in August 1982. It is by now little more than a gutted shell, any re-usable parts having been salvaged.

A strange visitor to Northfields Depot on the same occasion is this four-car unit of Metropolitan Line A Stock. Of note is how the leading driving motor car has had its original London Transport fleet name replaced by an Underground roundel, and it appears to be devoid of fleet numbers!

A private visit to Ruislip Depot, also in November 1990, reveals the tunnel cleaning train (Cars TCC1-TCC5) to have been resurrected from its sojourn in the sidings at the rear of Ealing Common Depot, and also freshly repainted, over the inspection pits.

Also fresh from overhaul in Ruislip Depot that day is this unit of 1962 Stock, headed by driving motor car 1536, looking very smart indeed.

To show to what extent graffiti can reach if rolling stock is left idle and unattended long enough, the Standard Stock driving motor car illustrated (L131) just a few months previously in Ruislip Depot is almost covered.

***Above*: Accompanying the** two vandalised Standard Stock pilot motor cars is the unique Acton Works shunter L11, formed from Standard Stock driving motor cars 3080 and 3109 in 1964; fortunately it was rescued for preservation. Sadly, however, its partner L130, once 3690, visible on the right had been saved too, but was scrapped in 2013 owing to its poor condition.

***Opposite above*: Another sad** sight in the yard of Ruislip Depot that day is this 1962 Stock trailer, which appears to be cannibalised for spare parts following accident damage. These cars would all be withdrawn within less than five years, other than a few moved to the Northern Line which lasted until the turn of the present century.

***Opposite below*: The former** Isle of Wight Standard Stock cars have been assembled in the sidings at Ruislip Depot, amid hopes that a 'heritage train' of them could be formed for operations on tours and so on. Alas, they were generally found to be unfit, notably owing to damage to underframes, etc. caused by sea water when travelling along Ryde Pier. Lack of funding played its part in this project being abandoned, too, and open storage at Acton didn't help, either. An interesting comparison is provided here with a train of 1962 Stock, which of course had replaced Standard Stock on the Central Line in the early 1960s.

RydeRail
Network SouthEast
10

***Above*: Not long** afterwards, a weekend celebration marked the Northern Line's centenary. The 1959 Stock train (one three-car and one four-car unit) given special treatment in 1920s livery operated a tour of the Northern Line complex, notably a trip around the supposedly haunted Kennington loop! This view finds it heading south at East Finchley on the Saturday.

***Opposite above*: Also at** East Finchley that day, a notable survivor and, apart from the 1938 Stock trailers working with 1960 Cravens driving motor cars on the Central Line, the oldest passenger stock still regularly working on the Underground, driving motor car 1000 and its unit of 1956 Prototype Stock are at the trailing end of a train bound for Kennington via Charing Cross. This was the first unpainted aluminium tube stock and, sadly, despite pleas to preserve at least one car of it, it was scrapped a few years later.

***Opposite below*: A relic** of the Northern Line's original ancestor, the C&SLR, was Angel Station at this time. Here, a train of 1959 Stock which has come from Mill Hill East heads south for Morden. The narrow island platform could be quite frightening for passengers at busy times, particularly if two trains arrived together when the platform was crowded during the rush hour. The same was true late at night after a few pints too many, as I well remember from my regular visits to the nearby King Edward VI in those days! Two years later, the station was drastically rebuilt, with a new northbound platform created and the existing ones being used for southbound trains, with of course a dividing wall and cross-passages between them. However, Clapham Common and Clapham North Stations retain this antiquated arrangement to this day.

MILL HILL EAST

***Above*: A ride** along the original path of the C&SLR (apart of course from the section between Borough and King William Street) on a train of 1959 Stock took me to Morden, where there was an open day at the Northern Line depot. Since this is some distance south, beyond the terminus, the by-now preserved unit of 1938 Stock ('topped and tailed' by driving motor cars 10012 and 11012) shuttled between the station and the depot, as illustrated here. It had been withdrawn from normal service some two and a half years earlier, the other driving motor cars that survived until that time with it having started new lives 'overseas' on the Isle of Wight.

***Opposite above*: Some of** the Standard Stock cars 'repatriated' from the Isle of Wight, initially taken to Ruislip Depot, were displayed in Morden Depot, two in Network SouthEast livery, and the two repainted in 1920s style. Such cars, of course, would have been based at Morden Depot when both they and the depot were new.

***Opposite below*: Pride of** place, to me anyway, amongst the exhibits at Morden Depot went to the Standard Stock control trailer restored to 1920s livery. This had been one of the last such cars in use with London Transport, running literally beneath my home in Canonbury on the Northern City Line until the autumn of 1966. At the time of writing, it still survives, awaiting full restoration at the LT Museum's Acton Depot.

485 044
RydeRail

6

***Above*: In total** contrast to the Standard Stock cars, this unit of Victoria Line 1967 Stock, with driving motor car 3110 leading, has been repainted in a different experimental livery to that of the similar 1972 Mk1 Stock train unveiled at East Finchley in July 1990, with the upper part of the body in dark blue and the lower in white. It remained unique in this style, its fellows receiving the latter's livery a few years later.

***Opposite above*: As is** customary at depot open days, one of the Underground's 'maids of all work', a battery locomotive is on show to the public, freshly repainted. It is not displaying its fleet number in the usual position in the centre of the cab, but is in fact L44, which also has its sides painted green as a departure from usual practice. It is the first of the eleven such locos supplied by British Rail in 1974, the last to perpetuate the general body design which first appeared in the mid-1930s.

***Opposite below*: In Morden** depot's open sidings, 1959 Stock driving motor cars 1311 and 1159 head trains stabled on the Saturday of the weekend it was open to the public. They are typical of the Northern Line's rolling stock at this time, some of which will remain in use until finally replaced in the first few weeks of the present century.

10
11
10
DANGER
620

MORDEN
via Bank
072

***Above*: On the** second day of the festivities, the train of 1959 Stock painted in an approximation of 1920s livery has disgorged its load of (mainly) transport enthusiasts onto the single platform of Mill Hill East Station. This isolated, single-track stub on the line which should have continued to Edgware and onwards to Bushey Heath as a double-track railway, has often presented operational difficulties for London Underground. One notable example was, shortly after OPO on 1995 Stock was introduced, a (now deceased) friend of mine who was driving one of them collapsed when changing ends at Mill Hill East. This, of course, caused the service there to be suspended for a considerable time while arrangements were made to rescue him – and his train.

***Opposite above*: The Northern** Line's 1959 Stock trains had been displaced to it by new 1973 Stock on the Piccadilly Line between 1975 and 1978. One of these, headed by driving motor car 166, arrives at Hammersmith Station in September 1991, heading east and bound for Arnos Grove. The station had been rebuilt at track level, in typical Charles Holden style, in conjunction with the Piccadilly Line's westward extensions nearly sixty years previously, requiring four-tracking, the existing District Railway (as it was then) using the outer two tracks.

***Right*: From the** late 1980s onwards, despite the Thatcher government's apparent disdain for public transport which was therefore starved of funding, a considerable amount of work was done refurbishing deep-level tube stations. This was especially so in Central London, and also involved renewing signage. In April 1992, a curious survivor on the Northern Line's northbound platform at London Bridge Station is this sign still showing the Northern City Line between Moorgate and Drayton Park as being part of the Northern Line, despite its having closed to Underground trains in October 1975! It dates from the time of the Victoria Line's opening in 1968/69 and had replaced a sign still showing the uncompleted Northern Line extensions in full. Very many Northern Line stations still had signs showing these until the advent of the Victoria Line, with the 'offending' parts plated over (or covered with stickers), so as not to confuse passengers and, perhaps, draw attention to the wasted works and public expenditure involved.

Left**: An even** odder survivor at the same period is this battered sign at the foot of the escalators in the Bakerloo Line's Warwick Avenue Station. A very close look shows that an additional station is shown beyond its traditional southern terminus of Elephant & Castle – Camberwell. Powers had been obtained to extend the line to that point, and it actually appeared on tube maps at the end of the 1940s, before the idea was placed in abeyance, if not abandoned completely. Both tunnels at Elephant do extend some distance south of the station, sometimes used to stable trains, but that is as far as the extension got – as far as we know. Today, plans to extend the line southwards, albeit on a different alignment, are once again in abeyance due to lack of funding.

Opposite above**: The much-needed** rebuilding of the Northern Line's antiquated Angel Station was completed in the autumn of 1992. The entrance was re-sited from City Road around the corner into Islington High Street, more convenient for shops and businesses and nearby Chapel Market. Its lifts were replaced by two banks of escalators, which are the longest in this country and also the fourth longest in Europe. Here, in September 1992, reopening day approaches.

Opposite below**: The Diamond** Jubilee of the Piccadilly Line's northern extensions was celebrated, also in September 1992, by an exhibition in the booking hall of Arnos Grove Station, and special runs between there and Cockfosters. Driving motor car 135 heads a specially spruced-up train of 1973 Stock, complete with commemorative headboard.

TATION
ANGEL STATION
UNDERGROUND

COCKFOSTERS
Piccadilly Line
1932
1992
Diamond Anniversary
101

***Above*: The train** of 1959 Stock which had been specially painted for the Northern Line centenary two years previously gave rides for this event too. Also bearing the special headboard, it had of course previously operated on the Piccadilly Line. The young lady standing beside it is my elder daughter Margaret, who accompanied me on the trip to Cockfosters and back.

***Opposite above*: A crowd** of enthusiasts throng the platform at Arnos Grove as the 1959 Stock train arrives to take another return trip. Regrettably, this smart unit too has been the target of graffiti taggers, hence the smudges on the cab front.

***Opposite below*: In January** 1993, this sign in the circulating area at the foot of the escalators at the Piccadilly Line's Turnpike Lane Station has apparently been uncovered, revealing the original name for Oakwood Station, which was called Enfield West until 1946. It dates from when the station opened in 1932.

ARNOS GROVE

SOUTHGATE
COCKFOSTERS
WOOD GREEN
BOUNDS GREEN
ARNOS GROVE
SOUTHGATE
ENFIELD WEST
COCKFOSTERS

***Above left*: Other** interesting signage on the Piccadilly Line still in evidence at this period was at Arsenal Station, notably its original solid roundel station names, dating from when the station opened in December 1906. The name plate dates from 1932, when the station was renamed from Gillespie Road. The suffix 'Highbury Hill' (whose northern end the station entrance faces) was later dropped. Following refurbishment, some such signs here were retained as also at neighbouring Holloway Road and Caledonian Road.

***Above right*: This** sign for Arsenal's eastbound platform dates from when the Piccadilly Line was extended north of Finsbury Park in 1932; the station was also renamed at that time. A prominent enamel plate reading 'Oakwood' is fixed above the original 'Enfield West', the plate's length is explained by the fact that the suffix 'Oakwood' was added to the original name a year or so after Enfield West opened. This sign was replaced upon refurbishment of Arsenal's platform area. The crookedly placed 'No Smoking' sign did not do it justice.

***Opposite above*: A common** feature of the original sections of the Bakerloo, Northern ('Hampstead tube') and Piccadilly Lines opened in the first decade of the twentieth century was that each had the station name prominently incorporated in the glazed tiling on their platforms. Despite the name being changed more than ninety years ago, this heritage feature has been retained at Arsenal Station. These three pictures date from February 1993.

***Opposite below*: A similar** tiled survivor is that at Hampstead Station on the Northern Line, pictured in March 1993. It was originally meant to be called Heath Street, being located at the junction of that thoroughfare and Hampstead High Street, thus the name was incorporated in the tiling. It was changed, however, before it opened in 1907. The station is London Underground's deepest, its platforms being accessed by lifts.

ARSENAL
ARSENAL
ARSENAL
ARSENAL
GILLESPIE ROAD

AMPSTEAD
HAMPSTEAD
HEATH · STREET

***Above*: At the** spring bank holiday weekend 1993 'Steam on the Met.' event, ex-GWR 5700 Class 0-6-0PT L99 bears the livery it gained upon acquisition by London Transport from British Railways in 1963. It was originally numbered 7715 and new in 1930. It approaches Chorleywood Station, hauling a former BR Southern Region 4TC unit, also adorned in maroon livery for use on such occasions. LT's pannier tanks, of course, were only usually used for engineers' trains or, more famously, to bring rubbish collected from around the system to Watford tip sidings until their final withdrawal in 1971.

***Opposite above*: Another splendid** original station sign still in place, pictured on 28 May 1993, is that on the Central Line's Bank Station eastbound platform. Dating from before the war and featuring the eastern extensions beyond Liverpool Street then due to open in the early 1940s, but not fully completed until 1957, it has several interesting features. The line to Ongar is shown as a continuous line, and not as a shuttle to and from Epping as it always was after the Central Line took over. Neither is the Woodford to Hainault section shown as a shuttle, as it was for many years and at the time I took this picture. The diagram implies that trains ran to Hainault either via Newbury Park (as they did for many years) or via Woodford, which was not usual. Today, the stations between Hainault and Woodford are usually served by trains running via Newbury Park to Hainault and continuing to Woodford. In addition, the sign originally showed Chigwell Lane Station, which was renamed Debden when electric trains finally reached it in 1949. This sign was replaced upon Bank Station's refurbishment.

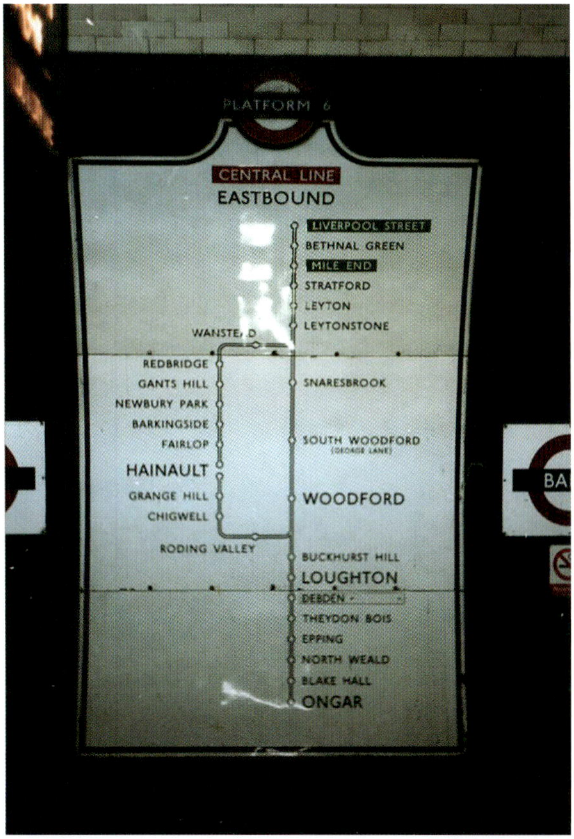

***Below*: It is** common practice for buses, traditionally those of London Transport, its predecessors and successors, to provide replacement services for Underground lines when closed for any reason. On 5 July 1993, LRT subsidiary London Northern has provided one of its private hire fleet, DMS2168 (one of a 'dying breed') as a replacement for the Northern Line's Finchley Central to Mill Hill East shuttle. Surely such a large vehicle is an over-provision for this short journey? Here, it heads along Dollis Road, Church End, having just passed beneath the Dollis Brook Viaduct, which carries the single-track branch and is the highest point above ground level (almost 60ft) that London's Underground reaches.

***Above*: Star of** the show at an open day at Ealing Common Depot on 25 September 1993 is former District Railway G Class (Q23 in LT terminology) driving motor car 4248. It is coupled to a Q35 trailer and Q38 driving motor car, and contrasts with the graffiti-sullied D Stock car on the adjacent track. These three Q Stock cars survive today, but 4248 is now a static exhibit in the LT Museum.

***Opposite above*: At the** other end of the formation is the Q38 driving motor car, one of a pair that has been preserved, 4416 and 4417 which had latterly been pilot motor cars L126 and L127, used for depot shunting. There are hopes that these may, someday, be restored to working order by the LT Museum and used for tours, in a similar fashion to the preserved 1938 Stock tube unit.

***Opposite below*: Also on** show that day is preserved R49 Stock driving motor car 22679, one of only six of that type. It has unpainted lightweight aluminium bodywork, whereas all other R Stock driving motor cars were R38s converted from Q38 trailers which were originally red and had steel bodywork. Unlike the older cars, R47, R49 and R59 cars (all being non-driving motor cars otherwise) had two large windows amidships rather than four smaller ones on the R38s.

049
22679

Current rolling stock was also on display at Ealing Common depot, but this uncoupling non-driving motor car of D Stock, detached from its unit, does not make a very edifying sight, with the ghosts of graffiti damage evident on its lower panels. Of note are its opening side windows, as added a few years after they first entered service in 1980.

Even less pleasing to the eye is this unfortunate diesel shunter, liberally marked with graffiti. It is DL81, one of three Rolls-Royce engined 0-6-0 shunters acquired second hand by London Transport in 1971 for use at Lillie Bridge and Neasden Depots, ostensibly to replace steam engines. They had been new in 1967/68 for use at an iron ore mine in Corby. For a variety of reasons, they saw little use on the Underground, and had been withdrawn by 1993. However, all three were subsequently preserved on 'heritage' railways.

The venerable tube gauging car, G663, also appears to have been 'put out to grass' at Ealing Common by this time. It had been converted from Standard Stock trailer 7131 in 1963.

By March 1994, new units of 1992 Stock, based on the 1986 Prototype Stock illustrated earlier, were well on their way to replacing the Central Line's 1962 Stock trains. Representing the old order, driving motor car 1434, now bearing much larger car numbers, perhaps to deter the graffiti vandals such as those who have defaced its cab front, heads a West Ruislip-bound train into Leytonstone Station.

***Above*: Arnos Grove** Station, opened in 1932 for the Piccadilly Line's northern extension to Cockfosters, is one of the most distinctive and iconic Charles Holden designs for Underground Stations from that period. This view on 8 May 1994 showing its surface buildings is, however, somewhat marred by the lower half of the 'bullseye' circle on the Underground sign being missing, and one of the bus stops (used for buses terminating in its forecourt) looking as if it is about to fall over.

***Opposite above*: Illustrating how** careworn and 'battle scarred' thanks to graffiti attacks they had become by this time, Metropolitan Line A62 Stock driving motor car 5136 arrives at Rickmansworth Station on 22 May 1994 heading an Amersham train. Perhaps their refurbishment had been deferred owing to proposals for the Amersham and Chesham services to become one of Crossrail's western branches – but, of course, it never happened after all.

***Opposite below*: A sight** that will be familiar to many Underground staff is the mock-up 'Wood Lane' station within their training centre at White City, used to familiarise them with various aspects of operation. Situated adjacent to the Central Line facilities there, it was later relocated to West Kensington. I took this view in September 1994, a few years before that happened.

FIRE EXIT
WOOD LANE

***Left*: The little-used** Piccadilly Line Aldwych branch finally closed to passengers at the end of September 1994. This view shows the sign on the wall of the sole platform that had been in use for most of the station's life, leaving no doubt as to where the trains went. The panel showing 'and Central Line' was presumably added when Holborn Station was rebuilt in 1932 with additional platforms for that line, replacing the nearby British Museum Station.

***Below*: Aldwych Station** had originally been named Strand, as this tiling clearly shows. It had been opened in 1907 and was renamed eight years later. As is well known, Aldwych Station and the disused second tunnel to Holborn were used in both world wars both as air raid shelters, and to store precious artefacts from nearby museums and art galleries. Both before and after closure, the station and short branch have been used for filming, notably the gruesome horror films *Death Line* (1972, dir. Gary Sherman) and *Creep* (2004, dir. Christopher Smith), both perhaps guaranteed to frighten people away from the Underground for evermore!

A much more modern London Underground sign that will soon be no more is that heralding the remote Ongar terminus of the Central Line. The shuttle service to and from Epping serving it closed too at the end of September 1994, when I took this photograph. Of note are the typical Great Eastern Railway station buildings – ironically, had this branch not been taken over by London Transport, it may well have been closed during the Beeching area thirty years previously – if not before!

On the penultimate day of Underground services to Ongar, 29 September 1994, 1938 Stock trailer 4927 is in the centre of the three-car shuttle, 'topped and tailed' by Cravens 1960 Stock driving motor cars when calling at North Weald Station. Most passengers aboard are probably railway enthusiasts.

The three-car unit looks somewhat lost in North Weald Station, as it prepares to set off for Ongar. Of note is the signalbox on the disused platform. It had been boarded up when the second platform was taken out of use but has now been restored in anticipation of a preservation group taking over the branch as a 'heritage' railway. The line is still in such use at the time of writing.

On the final day of the Ongar service, 30 September 1994, the last train to serve it, also a 1960 Stock unit with 1938 trailer, has run out from Hainault depot to Woodford Station, and awaits being able to shunt into the sidings, and reverse onto the eastbound track as a unit of 1992 stock comes out of the sidings ready to take up service.

Meanwhile, an eight-car train of 1962 Stock departs from Woodford Station bound for West Ruislip, with driving motor car 1516 at its trailing end. By now, these were rapidly being replaced by new 1992 Stock units, which proved to be nowhere near as reliable the old trains. They had taken over completely by the summer of 1995.

Dusk falls and enthusiasts on Ongar Station crowd aboard the last Underground train to serve it. The unit, comprising cars 3906, 4927 and 3927, was subsequently privately preserved and has since carried out several tours on various Underground lines. It is a great credit to those who care for it.

***Above*: On 20 April** 1995, an eight car train of Metropolitan Line A Stock speeds past West Hampstead Station on its way to Central London. By now, these reliable units had seen best part of a quarter of a century of service and were sorely in need of external tidying up. The station itself had new platform buildings for the Bakerloo Line's takeover of the stopping service between Baker Street and Wembley Park in 1939, with the original Metropolitan Railway booking office on the overbridge retained. These are visible on the left.

***Opposite above*: A surprise** on the same occasion is the appearance of Metropolitan electric locomotive 12 *Sarah Siddons* with a brake van, probably one of those of traditional design built for London Transport by British Railways as recently as the 1960s, and a battery locomotive in tow. *Sarah Siddons*, of course, was and still is active on railtours on the Underground.

***Opposite below*: Also that** day, 1983 Stock driving motor car 3732 is at the trailing end of a northbound Jubilee Line train departing from Finchley Road Station. Despite about half of these trains being delivered only in 1988, including this one, all were replaced by new 1996 Stock by the summer of 1998, most eventually going for scrap and none seeing further operational use.

12
465

CHARING CROSS
T

***Above*: In contrast** to the short-lived 1983 Stock trains, 1959 Stock still formed the backbone of the Northern Line fleet at this time. One unit had been given the new livery adopted as standard with a view to refurbishing them for a few more years' service, and on 16 May 1995 arrives at Finchley Central Station heading a High Barnet train.

***Opposite above*: Driving motor** car 1043 and trailer 2042 are at its leading end on this northbound service at Finchley Central, and the new livery suits its well-tried design, derived of course from 1938 Stock. However, by now new 1995 Stock had been ordered to replace the entire Northern Line fleet. These older units, therefore, were not refurbished after all.

***Opposite below*: On the** same occasion, one of the 1972 Mk1 Stock units given the new livery in 1990 is coupled to one still in unpainted condition when at the trailing end of a Mill Hill East-bound service.

1043
2042

MILL HILL EAST

***Above*: Graffiti mars** the cab of 1959 Stock driving motor car 1252, at the trailing end of a southbound train at Finchley Central that day. It is bizarre that it shows the incorrect destination of Edgware via Bank, since it is not possible to travel directly by Northern Line from Edgware to Finchley Central! But, of course, it would have been if the extension from Mill Hill East to Edgware had been completed...if only!

***Opposite above*: Headed by** driving motor car 1079, another graffiti-ridden train of 1959 Stock arrives at Finchley Central bound for Mill Hill East. Of note is the new brick-built waiting room on the left, the only major alteration to this traditional station. It has recently replaced the old wooden GNR structure.

***Opposite below*: The same** train has arrived at Mill Hill East, beyond which a short stretch of non-electrified track disappears into the foliage. At this period, serious consideration had been made to re-extending the line from here to serve Copthall Sports Centre, which is alongside the line, about half way towards Mill Hill Hale Station's site. Alas, it was not to be, and in recent years residential development has taken place directly on this alignment.

0 3 2

1079

***Above*: On 20 May** 1995, a train of A Stock, whose reasonably smart appearance has been wrecked by graffiti on the cab front, has paused briefly at the southern end of Rickmansworth Station, perhaps in conjunction with the 'Steam On The Met.' event in progress that weekend. The preserved British Rail Eastern Region Class 306 'Shenfield Slider' EMU is on display in the bay platform.

***Opposite above*: Pictured from** the steps of the old signalbox at the southern end of the station, a relatively unblemished train of A Stock calls at Rickmansworth on its way to Aldgate. A newer structure, dating from the time the line was electrified beyond this station is visible on the left at the far end of the northbound platform. The cumbersome-looking structures on the right are CCTV cameras to allow drivers to see all along the curved platform that passengers are clear of the doors before closing them – all a far cry from when electric locomotives replaced steam at this point prior to the 1961 electrification to Amersham!

***Opposite below*: On the** same occasion, two trains of A Stock, with A60 driving motor car 5006 on the left and A62 driving motor car 5189 on the right leading, show just how thoroughly graffiti vandals have defaced them when stabled in sidings south of Rickmansworth Station. Of note is how their car numbers have now been placed above their windows, no doubt to avoid being obscured.

ALDGATE

CHESHAM
METROPOLITAN
5189

***Above*: Upon completion** of the Metropolitan main line electrification, trains terminated at Amersham, abandoning the section beyond to Aylesbury to British Railways trains to and from Marylebone and the former Great Central main line. Sadly, the latter was to be axed in a belated implementation of the Beeching cuts north of Aylesbury in 1966, and the entire line was threatened with closure during the Thatcher years twenty or so years later. Fortunately, it survived, and a thriving service between Marylebone and the West Midlands, albeit via the former Great Central and Great Western route, operates today. Here, a fast Metropolitan Line service, which skips intermediate stations such as Northwick Park and Northwood Hills, terminates at Amersham. Of note is how the leading driving motor car's 'London Transport' fleet name has faded away almost to nothing, thanks to repeated cleaning. Application of London Underground roundels was never completed on all such A Stock prior to their refurbishment.

***Opposite above*: Graffiti has** even spread to the ill-fated Highgate High Level Station. In this view on 8 July 1995, foliage has again engulfed the southern end of the platform. In an attempt to prevent water seepage to the booking hall and escalator shaft below, plastic corrugated sheeting has been placed above the trackbed. A gangway above this from the station's rear entrance has been provided for staff access. The station is now securely fenced off, as are the tunnels at each end.

***Opposite below*: On the** same day, a Saturday, Northern Line services are terminated from the south at East Finchley, trains having to reverse north of the station. One such, a train of the then-still dominant 1959 Stock, has arrived in the northbound platform and is changing crew.

EAST FINCHLEY
2 5
3/91

***Above*: Taking up** service on the southbound platform another train of 1959 Stock collects passengers for its next trip below Central London, in this case on the Charing Cross branch to Kennington. Its guard keeps a watchful eye before closing his doors. Following the conversion to OPO of the Central Line and withdrawal of its 1962 Stock, the Northern Line was the only Underground line to retain guards. Of note here is the 'Archer' symbol below the car's last overhaul date on the left, taken from the statue and the southern end of East Finchley Station and indicating it belongs here.

***Opposite above*: The reason** for the curtailment of Northern Line services at East Finchley was engineering works on the section between that station and Finchley Central, to where trains from High Barnet and Mill Hill East were likewise curtailed. Here, another train of 1959 Stock departs from the first station north of that point, the quaint West Finchley.

***Opposite below*: Showing to** good effect some of the ancient station buildings retrieved from closed East Anglian branch lines when the LNER built West Finchley Station in 1932, a train of 1972 Mk1 Stock arrives there bound for Finchley Central. Such units were becoming rare on the Northern Line by this time. Having been cobbled together as a hastily improvised replacement for ailing 1938 Stock, and effectively a crew-operated version of the Victoria Line's 1967 Stock, they suffered from various defects and were not liked by crews. Several cars of the type were 'robbed' to bolster services on the Victoria Line and the Bakerloo Line, where they were modified to work with 1967 and 1972 Mk2 Stock respectively. Others were cannibalised for spare parts. None survived until the very end of crew operation on the Northern Line, the last having been withdrawn a few weeks beforehand.

FINCHLEY CEN
105

FINCHLEY CEN

On the same day, another train of Northern Line 1959 Stock terminating at Finchley Central departs from Totteridge & Whetstone Station. This station has Great Northern Railway architecture typical not only of the Edgware and High Barnet branches, but also of their Enfield branch opened in 1870; in fact, this station is virtually identical to that at Palmers Green, save that the latter's canopied Up platform building is longer and not divided into two sections as this one is.

By May 1996, the Metropolitan Line's long-suffering A Stock trains are finally being refurbished, and very nice they look in the Underground's new livery. Since following OPO conversion ten years previously, when not all driving motor cars were equipped for that, trains ran in semi-permanent eight car formations. Those driving motor cars not so equipped did not receive red cab fronts, as A60 driving motor 5039 in the sidings south of Rickmansworth Station shows.

On a damp and dreary January 1997 day, Northern Line services are again curtailed at East Finchley, this time from the north, owing to engineering works. Two trains of 1959 Stock stand in the centre platforms meant for trains coming from the Northern City Line via Finsbury Park, the one on the left due to run to Mill Hill East. Just three years later, scenes like this would be history, with the line worked entirely by new 1995 Stock.

In May 1997, one of the units of Northern Line 1972 Mk1 Stock painted in the new livery has been paired with another that has not. It calls at Finchley Central Station on its way to Morden via Bank.

***Above*: At the** 1997 August Bank Holiday weekend, the preserved unit of Cravens 1960 Stock (still with 1938 trailer) worked specially between Chesham and Chalfont & Latimer in place of the usual train of Metropolitan Line A Stock. This view taken through the trees south of Chesham Station finds it departing from that terminus.

***Opposite above*: During the** mid-1990s, the East London Line had to be closed owing to necessary repairs to Brunel's famous Thames Tunnel between its Wapping and Rotherhithe Stations. At the same period, plans were agreed to extend it at both ends, most notably to Dalston Junction in the north, with a link to the former North London Line (which the Thatcher government had been responsible for closing in 1986), necessitating the closure of its Shoreditch terminus. This view of that establishment on August Bank Holiday Monday 1997, taken from the former Bishopsgate Goods Station, shows it looking very neglected. However, in the event, it did reopen and remained in use until June 2006, the extension and connection to the NLL finally being opened in May 2010, but under London Overground control and no longer part of the Underground network.

***Opposite below*: The eastward** extension of the Jubilee Line, well on the way to completion by now, spelt the end of the short-lived 1983 Stock. On 28 March 1998, driving motor car 3735 heads a train bound for Charing Cross, which terminus would also be abandoned, at West Hampstead Station. These units would all be withdrawn a few weeks later, and after several years in store and at the mercy of graffiti vandals, most were scrapped.

CHARING CROSS
3735

***Above*: Looking very** smart indeed, a train of refurbished Metropolitan Line A Stock calls at Chalfont & Latimer Station in May 1998, with A62 Stock driving motor car 5225 at its trailing end. The bay platform for shuttle services to and from Chesham is on the left.

***Opposite above*: By now** 'Steam on the Met.' events had become a regular feature, taking place over the Spring Bank Holiday weekends at the end of May. Latterly, displays of Underground rolling stock were included in the sidings south of Rickmansworth Station, where one of the exhibits at the 1998 event is this unit of Circle/Hammersmith & City Line C69 Stock, with driving motor car 5537 facing the camera, adorned as an all-over advertisement for Yellow Pages. Such advertising was relatively new to the Underground but had featured on London buses since the late 1960s, and by coincidence, one of the first so treated was for the same advertiser.

***Opposite below*: Another exhibit** that day is London Underground's track recording train, still with its two Cravens 1960 Stock driving motor cars with the 1973 Stock trailer in the middle. It has recently been repainted, and stands next to the preserved Cravens unit, with which it once ran on the Central Line when new.

WEMBLEY PARK
245
YELLOW PAGES
http://www.yell.co.uk
YELLOW PAGES
http://www.yell.co.uk
YELLOW PAGES
5098

466

On 13 June 1999, the preserved Cravens 1960 Stock unit operated a railtour on the Northern Line. This view finds it passing through Golders Green Station's platform two. The unit is a great credit to its owners and resides at Hainault Depot from where it had operated in service.

Sadly, 1959 Stock on the Northern Line was now in terminal decline, with new 1995 Stock rapidly replacing them. At Brent Cross Station, also on 13 June 1999, the three-car unit that had been adorned in 1920s livery in 1990 is now coupled to a four-car unit in the usual colours when heading north towards Edgware.

1959 Stock driving motor car 1292 heads a train bound for Morden via Bank when emerging from the short tunnel north of Hendon Central Station.

Over on the High Barnet branch, another train of 1959 Stock represents the old order when arriving at Woodside Park Station heading south. The station car park on the left appears to be empty on this Sunday and shares the site of the former goods yard with a recent housing development.

***Above*: The preserved** Cravens unit has, meanwhile, been to Edgware and back and now explores this branch, as it heads north from Woodside Park passing an approaching train of 1995 Stock. All too soon, the latter units would be taking over the entire Northern Line service.

***Opposite above*: Upon arrival** at High Barnet terminus, those riding on the tour are able to take a break, as does the Cravens unit and its crew. Despite plans having been made to build a new booking hall here with better access to the Great North Road above as part of the 1935-1940 New Works Programme (as well as earlier plans to extend the line in tunnel further north), this splendid Great Northern Railway Station remains in largely original condition. In recent years, however, a level walkway between the two platforms illustrated here has been provided, leading behind the buffer stops, to assist passengers with mobility problems and also obviate having to build lifts to the footbridge.

***Opposite below*: Commenced at** the end of 1993, work on the Jubilee Line extension took some six years to complete, and as seems to be endemic with such projects nowadays, was plagued with cost increases and work over-runs, and even after opening such things were still a problem, notably with signalling. The extension was opened in stages between May and December 1999. Its route was significantly different from that originally envisaged, perhaps owing to the machinations of International Finance, whereby one such institution based on the Isle of Dogs provided a substantial amount of funding for it, naturally ensuring it would serve that area. On 13 August 1999, 1996 Stock driving motor car 96041 has arrived at its new terminus in the East, Stratford. Basically similar to the Northern Line's 1995 Stock, a slight difference in frontal styling is evident when compared with the picture of one of those above.

SPECIAL
734

96041

Above**: On the** evening of 27 January 2000, history was made when the last two-person operated train in normal service on the Underground ran. This was Northern Line train 132 from Morden to Edgware and back, of which trailer 9257 is pictured here at Morden.

Opposite above**: Driving motor** car 1256 headed the train, seen here about to depart. This was, of course, also the last traditional tube train to run in normal service, as well as being the last 1959 Stock train in service, this type having achieved some forty years' service on the Piccadilly, Central, Bakerloo and Northern Lines.

Opposite below**: Some very** well-presented unofficial notices were produced by London Underground staff and posted in the cars on this historic last train. One of these may be read on the left in its trailing driving motor car during the journey.

EDGWARE
via Charing Cross

LAST "1959"
STOCK TRAIN
This is the
last old-style
NORTHERN LINE
train, and the final
Guard-operated
service on the
UNDERGROUND

***Above*: A final** look at the last 1959 Stock train, when calling at Hampstead Station on its last journey, taken from the leading car. A comparison between this and the earlier picture of this station, complete with the 'Heath Street' tiling, will show that the station has since been refurbished.

***Opposite above*: One of** the exhibits at the 2000 'Steam on the Met.' event was a train of recently-refurbished Piccadilly Line 1973 Stock. Here, on the May Bank Holiday Monday, it passes through Rickmansworth Station after reversing, and will proceed to Harrow on the Hill, where it will reverse again to gain the Uxbridge branch, from which it can return to its native line.

***Opposite below*: A relic** of the Piccadilly Line's early years remains at Hyde Park Corner Station in the cross-passage at the foot of the stairway from the station's original lifts. Such reminders of the line's past survive at most, if not all, of the stations on its deep-level section between Earls Court and Arsenal.

Piccadilly line
5042

TO THE TRAINS
TO HAMMERSMITH
TO THE TRAINS
TO FINSBURY PARK

***Above*: On an** early spring day, 18 March 2005, evening sunshine catches a train of Bakerloo Line 1972 Mk2 Stock as it approaches Willesden Junction 'new' station on a northbound service. No one could then have foreseen that these trains would still be running twenty years later – if not longer. I took it from the North London Line high level platforms; in the background is a Class 313 EMU, which at the time worked that line as well as the Euston to Watford DC services, sharing tracks with Bakerloo Line trains between Queens Park and Harrow & Wealdstone. Their depot is on the right of the picture.

***Opposite above*: Another C69 Stock** unit to gain all-over advertising promoted London's bid to host the 2012 Olympic Games, which of course it won. On 27 June 2005, driving motor car 5570 is at its trailing end on a District Line Edgware Road to Wimbledon service calling at Putney Bridge Station. Sadly, some cars of this type were destroyed in the terrorist attacks near Edgware Road and Aldgate/Liverpool Street Stations just ten days later.

***Opposite below*: On 20 April** 2006, District Line D78 Stock driving motor car 7076 heads a Richmond-bound train arriving at West Ham Station. This station had been substantially rebuilt in connection with the new Jubilee Line extension passing below in the late 1990s, and the main line London, Tilbury & Southend Line platforms on the right, which had been removed following serious damage in the 1940 blitz, reinstated.

5570
EDGWARE ROAD
076

5709
357 006

SHOREDITCH STATION
UNDERGROUND

***Opposite above*: Also at** West Ham, C77 Stock driving motor car 5709 calls with a Hammersmith & City Line service from Barking to Hammersmith Met. on 9 June 2006, as one of then recent Class 357 EMUs heads east on the main line platforms.

***Opposite below and right*: 9 June 2006** was also the last day of the old Shoreditch terminus of the East London Line, long since only used during Monday to Friday rush hours, and for the nearby Sunday morning markets. The modern Underground sign contrasts with the old building, whilst a prominent notice on the wall warns of the closure. This was necessary in order to divert the line onto a new alignment linking with the former North London Line. A new Shoreditch High Street Station was built straddling the nearby High Street, with operation transferred to Transport for London's new London Overground entity, as related earlier.

***Below*: Brilliant summer** sunshine catches A60 Stock driving motor car 5102 arriving at the old Shoreditch Station on its final day, in the early part of the evening rush hour. The platform was crowded when the final train departed at the end of it!

***Above*: Not far** away from the previous pictures, some withdrawn 1983 Stock cars are perched on the disused part of the former North London Line viaduct between the old NLR Shoreditch Station (closed owing to wartime bombing and never reopened) and Broad Street and have found further use as offices and art studios.

***Opposite above*: Presenting a** more pleasant spectacle, the original facade of the Great Northern & City Line Highbury Station's booking hall, which had been closed in 1968 when the station's lower part was combined with the Victoria Line and the entrance arrangements combined with the British Rail North London Line station on the other side of Holloway Road, has recently been revealed to the public again having previously been hidden by advertisement hoardings. Sadly, graffiti vandals and fly-posters have already attacked it, but their efforts have been swiftly obliterated when I took this picture on 16 October 2006. Then, as still now, plans exist to reopen this entrance/exit for step-free access, reinstating the station's lifts.

***Opposite below*: On 22 December** 2007, driving motor car 7060 is at the trailing end of an eastbound District Line train departing from Bow Road Station. It is one of the last un-refurbished D Stock units: sadly, all would be withdrawn by the spring of 2017 though many have been resurrected in various modes for National Rail operation, most notably on the Isle of Wight.

HIGHBURY STATION
1904
ENTRANCE
EXIT
CORNER
C London
Shoreditch
Islington

out
← Bow Church
DLR Station
UPMINSTER
7060
7060
District line
017
District line

***Above*: This date** was also the last day of London Underground operation of the East London Line, before its closure for adaptation for its new role as part of the new London Overground network and linkage to National Rail systems at both ends. This close-up of A60 Stock trailer 6064 shows how when this stock worked the line in its later days, they carried dual line-branding for the East London Line and the Metropolitan Line, the latter supplying and maintaining them. Also evident are the end windows added to this stock upon refurbishment in the mid/late 1990s, and the safety barriers between cars to prevent errant passengers falling between them at stations.

***Opposite above*: On the** last day of London Underground operation on the East London Line, 22 December 2007, A60 driving motor car 5113 is at the trailing end of a Whitechapel-bound train which has just departed from Surrey Quays Station. The station had been renamed from its traditional name, Surrey Docks, when rebuilt in the 1980s.

***Opposite below*: Fellow A60** driving motor car 5059 is at the trailing end of this unit departing from New Cross Station on the final day. This has remained a terminus for the London Overground service, whilst of course at New Cross Gate, the line has been reconnected to the main line, and trains continue to Crystal Palace and West Croydon.

NEW CROSS
5113
ER.2
172

NEW CROSS GATE
5059
173

To illustrate how the curse of graffiti can still wreck this appearance of smartly painted Underground rolling stock, C77 driving motor car 5717 has had its cab front ruined when at the trailing end of a Hammersmith & City Line train at West Ham Station on 7 February 2008.

At Barking Station on the same day, District Line D78 Stock driving motor car 7070 at the trailing end of an Upminster service is one of three different operators' rolling stock in evidence. A London Overground Class 150 two-car DMU has terminated on the left on their service from Gospel Oak, whilst a Class 357 EMU working for C2C on the former LT&S main line is over on the right.

Moving over to North West London, A62 Stock driving motor car 5163 heads a southbound Metropolitan Line service speeding past Co. Kilburn Station on 3 May 2008. The southbound Jubilee Line platform and track are on the left of the picture.

On the same day, two trains of Bakerloo Line 1972 Mk1 Stock are evident in this picture at the southern end of Wembley Central Station. The West Coast main line is on the left of their tracks, and the scene is little changed fifteen years later.

***Above*: Also that** same day, 1972 Mk2 driving motor car 3243 is at the trailing end of this northbound Bakerloo Line train calling at North Wembley Station. Platform signage has recently been changed here from the rather bland former British Rail signs latterly adorned with 'Silverlink' logos to London Underground ones, as visible on the right. They are only temporary signs for now, permanent ones having been erected subsequently.

***Opposite above*: At nearby** Northwick Park Station that day, A60 Stock driving motor car 5014 is at the trailing end of a Metropolitan Line stopping service heading for Aldgate. Fast services that skip this station, usually coming from Amersham, use the track on the left. This view also shows all the paraphernalia on the platform needed for OPO and also CCTV security surveillance necessary in the twenty-first century.

***Opposite below*: Further south** on the Metropolitan Line, A62 Stock driving motor car 5151 speeds past a Jubilee Line 1996 Stock train when heading a fast Amersham to Baker Street service at Dollis Hill Station. This picture graphically illustrates the difference in size between tube stock and sub-surface stock operated by London Underground.

ALDGATE 5014
421

Southbound platform 2

***Above*: Also at** Dollis Hill Station, but on 30 June 2008, this time looking south, A60 Stock driving motor car 5107 is at the trailing end of a Metropolitan Line train bound for Baker Street. With their comfortable deep-upholstered transverse seats, these trains gave a very pleasant ride, especially when at speed.

***Opposite above*: Another example** of a tube-size train being dwarfed by surface stock is this view at Kenton Station on 9 August 2008, with 1972 Mk2 Stock driving motor car 3232 at the trailing end of a northbound Bakerloo Line service alongside London Overground three-car EMU 313117. This is unusually working a Watford Junction to Stratford service (via the closed Primrose Hill Station), as operated during long-term engineering works on the North London Line between Willesden Junction High Level and Camden Road Stations.

***Opposite below*: A taste** of things to come at a special exhibition *in* Euston Square (as opposed to at the nearby station of that name) in October 2009 is this mock-up of one of the new S Stock trains, intended to replace all London Underground sub-surface stock. Those for the Metropolitan Line, in eight-car formations, have some transverse seating, but unfortunately the seven car formations for the Hammersmith & City/Circle and District Line are entirely composed of cars with only peripheral seating.

Stratford
313117

EUSTON SQUARE

On 16 October 2008, D Stock driving motor car 7027 is at the trailing end of a District Line train bound for Wimbledon. It is departing from West Brompton Station and about to descend into the tunnel taking these tracks beneath the West London Line into Fulham Broadway Station.

This smart new structure is the entrance to the Hammersmith & City Line's new Wood Lane Station, seen on 30 October 2008 shortly after opening. It is close to the line's former White City Station, whose entrance was on the other side of the overbridge and which closed in 1959. The former Central Line Wood Lane Station was located just south of the bridge.

On the same occasion, C69 Stock driving motor car 5584 heads a Hammersmith-bound train at the new station. This was built in conjunction with the opening of the nearby Westfield Shopping Centre, much of which is on former Underground land.

On 29 March 2009, A62 Stock driving motor car 5165 is at the trailing end of a Metropolitan Line train departing from Barbican Station in the evening rush hour, as one of its fellows arrives there on a northbound service. New S Stock was now being built to replace these units, some of which achieved more than fifty years at London's service.

***Above*: Boston Manor** Station has an interesting blend of original District Railway platform buildings, and Chas Holden surface-level structures dating from the Piccadilly Line's 1932 extension to Hounslow West. On 20 July 2009, one of its present-day 1973 Stock trains arrives there bound for Heathrow, a scene little changed at the time of writing.

***Opposite above*: A few** days later, 1972 Mk2 driving motor car 3263 is at the trailing end of a northbound Bakerloo Line calling at Willesden Junction 'new' station. Of note are the new station buildings above the train and on the left, built under National Rail auspices; the picture is taken from the bridge leading to the high level North London Line platforms.

***Opposite below*: On 5 August** 2009, much excitement was caused by the operation of a tour using London Underground's preserved 1938 Stock four-car unit over the eastern and southern section of the Jubilee Line. Railway enthusiasts and perhaps also members of the general public scramble to photograph it at Stratford Station, where the tour began and ended, and where it makes an interesting comparison with one of the Jubilee Line's current 1996 Stock trains.

Bridge to Jubilee line
Platform 13

Here, the 1938 Stock unit passes through the new Jubilee Line station at West Ham, which had opened ten years previously. On the right are the former North London Line platforms opened in 1979 but closed in 2006. They are being renovated for use by a new branch of the Docklands Light Railway and reopened in 2011.

The tour train terminated at West Hampstead, from where after a layover in the reversing siding north of the station, it returned to Stratford, taking a detour to the Jubilee Line's abandoned Charing Cross terminus on the way. In the meantime, A60 Stock driving motor car 5072 speeds past West Hampstead Station heading an Uxbridge-bound service.

District Line D Stock driving motor car 7110 looks very smart when standing at the line's Ealing Broadway terminus on 31August 2009. The reflection of a train of Central Line 1992 Stock may be discerned in its windows, standing at that line's terminus there too.

Ealing Common Station is a typical Charles Holden structure, rebuilt for when the Piccadilly Line took over the District Railway branch to South Harrow and beyond in 1932. It seems to overwhelm this train of Piccadilly Line 1973 Stock heading for that branch, also on 31 August 2009.

***Above*: On the** same occasion, another District Line D Stock train arrives at Ealing Common Station, having shared tracks from Acton Town to that point with the Piccadilly Line. Driving motor car 7104 heads it, bound for Ealing Broadway, just one station further on.

***Opposite above*: Until replaced** by new S Stock in 2013/14, C69/77 Stock was the only type permitted to work the District Line service between Edgware Road and Wimbledon after all CO/CP and R Stock had been withdrawn, since D Stock trains were too long for the Earls Court to Edgware Road section. On 20 April 2010, C69 driving motor car 5531 is at the trailing end of a Wimbledon-bound service departing from West Brompton.

***Opposite below*: An oddity** on the Bakerloo Line is 1972 Stock driving motor car 3299 and its unit. It had been created to replace cars of 1972 Mk2 Stock damaged beyond repair in a collision, using salvaged 1972 Mk1 Stock. On 8 May 2010, it is at the trailing end of a Harrow & Wealdstone service calling at South Kenton Station. This odd unit remains in service at the time of writing.

5531
WIMBLEDON

3299
3299

***Above*: In the** original through platforms of Whitechapel Station on the following day, C69 Stock driving motor car 5524 heads a Barking to Hammersmith train on the Hammersmith & City Line. This station is unrecognisable today following rebuilding and the opening of the Elizabeth Line.

***Opposite above*: By this** time, Circle Line services have been revised to include journeys on the anti-clockwise ('inner rail') service continuing via the Hammersmith & City Line to Hammersmith Met., and then returning from there and gaining the Circle clockwise ('outer rail') service at Edgware Road. C69 driving motor car 5596 works one of the former journeys at Barbican Station, passing one of its fellows heading for Barking.

***Opposite below*: In a** scene that can still be witnessed today, 1972 Mk2 driving motor car 3252 is at the trailing end of a Bakerloo Line northbound service departing from Kensal Green Station in September 2010. The station has received TfL signage following the takeover of the Euston to Watford DC stopping service, which uses the same tracks, by London Overground just under three years previously. The tunnel here is only a short one; daylight may be seen at the north end of the southbound bore.

5596
Circle line
to Hammersmith
225

***Above*: Also in** September 2010, work is under way refurbishing and extending Farringdon Station, initially for the overdue 'Thameslink 2000' programme and then for Crossrail. C69 driving motor car 5504 arrives there on what is presumably a traditional Circle Line trip that will continue from Edgware Road to High Street Kensington, Gloucester Road and onwards around the circle proper.

***Opposite above*: The clock** is ticking now for the Metropolitan Line's efficient and faithful A Stock trains, which have served London for almost fifty years, having initially replaced steam-hauled and electric compartment stock on the main Amersham and Aylesbury services in 1960/61, and then F Stock and P Stock on its Uxbridge and Watford branches in 1962/63. Also in September 2010, A60 driving motor car 5088 heads a stopping service into North Harrow Station.

***Opposite below*: To prove** the point, driving motor car 21004 heads a train of brand new S8 Stock northwards on a test run into Harrow on the Hill Station on the same occasion. Within almost exactly two years, their task will be complete on the Metropolitan Line.

BAKER ST
5088
451

21004

***Above*: Representing the** old order at this station on the same occasion, A60 Stock driving motor car 5018 is at the trailing end of a northbound service, its cab front scarred by graffiti. This station is yet another Charles Holden 1935-1940 New Works Programme rebuild, in anticipation of the Metropolitan Line's four-tracking, and electrification north of Rickmansworth which, of course, was not completed until the early 1960s when the A Stock trains were new.

***Opposite above*: Further out** into 'Metroland', A60 Stock driving motor car 5084 arrives at Pinner Station on one of the Metropolitan Line's stopping services, heading the usual eight-car formation on 17 September 2010.

***Opposite below*: Stablemate A** Stock driving motor car 5052 arrives at Pinner on a northbound Watford service. This view shows how the station retains its original platform buildings on the southbound side but had new ones and a new footbridge as a result of the Metropolitan Line four-tracking works in the late 1950s/early 1960s.

5084
440

WATFORD
5052
445

Above**: At the** Metropolitan Line's Watford terminus, another A Stock train awaits its passengers for its southbound journey towards Central London. When it departs, A60 driving motor car 5115 will be at its trailing end.

Opposite above**: Bakerloo Line** trains once served Watford's main station, Watford Junction, but now venture no further north than a reversing siding just north of Harrow & Wealdstone Station of course. Also on 17 September 2010, 1972 Mk2 driving motor car 3252 heads a seven-car train into the station to take up service. Beside it on the right, where part of the platform has been removed, is where the two locomotives (45637 *Windward Islands* and 46202 *Princess Anne* heading the Down express) landed after striking the wrecked 46242 *City Of Glasgow* which had caused the initial collision in the October 1952 Harrow disaster. They also fouled the southbound Bakerloo Line, on which an approaching 1938 Stock train had a narrow escape thanks to the engines causing a short circuit.

Opposite below**: Driving motor** car 3552 is at the trailing end of this 1972 MK2 Stock train awaiting departure from Harrow & Wealdstone Station on the same occasion. The section of the station footbridge above it which crossed the adjacent fast tracks of the West Coast main line was knocked down when the double-headed down express struck the wreckage of the up local train and the offending 46242 *City Of Glasgow*, some commuters who were on the bridge at the time were among the fatalities.

30

Customer
Lift
3552
3552

***Above*: Harrow &** Wealdstone Station still retains its original LNWR street-level buildings on the up (southbound) side, and standard London Underground ticket barriers are visible inside the entrance on the left. The station has both Underground and National Rail emblems (since semi-fast main line trains stop there too) and has since had the London Overground emblem added for the Watford DC stopping services. Beneath the emblems is a memorial tablet to the 112 poor souls who perished in the disaster there on 8 October 1952.

***Opposite above*: In total** contrast, the former LNWR local station at Kensal Green had its booking office rebuilt in this unusual style in 1980. The National Rail services indicated by the former British Rail emblem are, of course, those operated by London Overground for whose Euston to Watford service the station was originally opened when it was electrified in 1916. It has never had platforms on the four other main line tracks. Having a more frequent Bakerloo Line service, it is managed by London Underground today. The Overground symbol had yet to be applied when I took this picture on 29 September 2011.

***Opposite below*: On 2 November** 2011, there must be an exhibition on at Olympia in view of the number of passengers awaiting this District Line train arriving there, with D Stock driving motor car 7082 leading. For many years, the District Line only served the station when such events were in progress, as it does again today.

Kensal Green
UNDERGROUND

District line
7082
7082

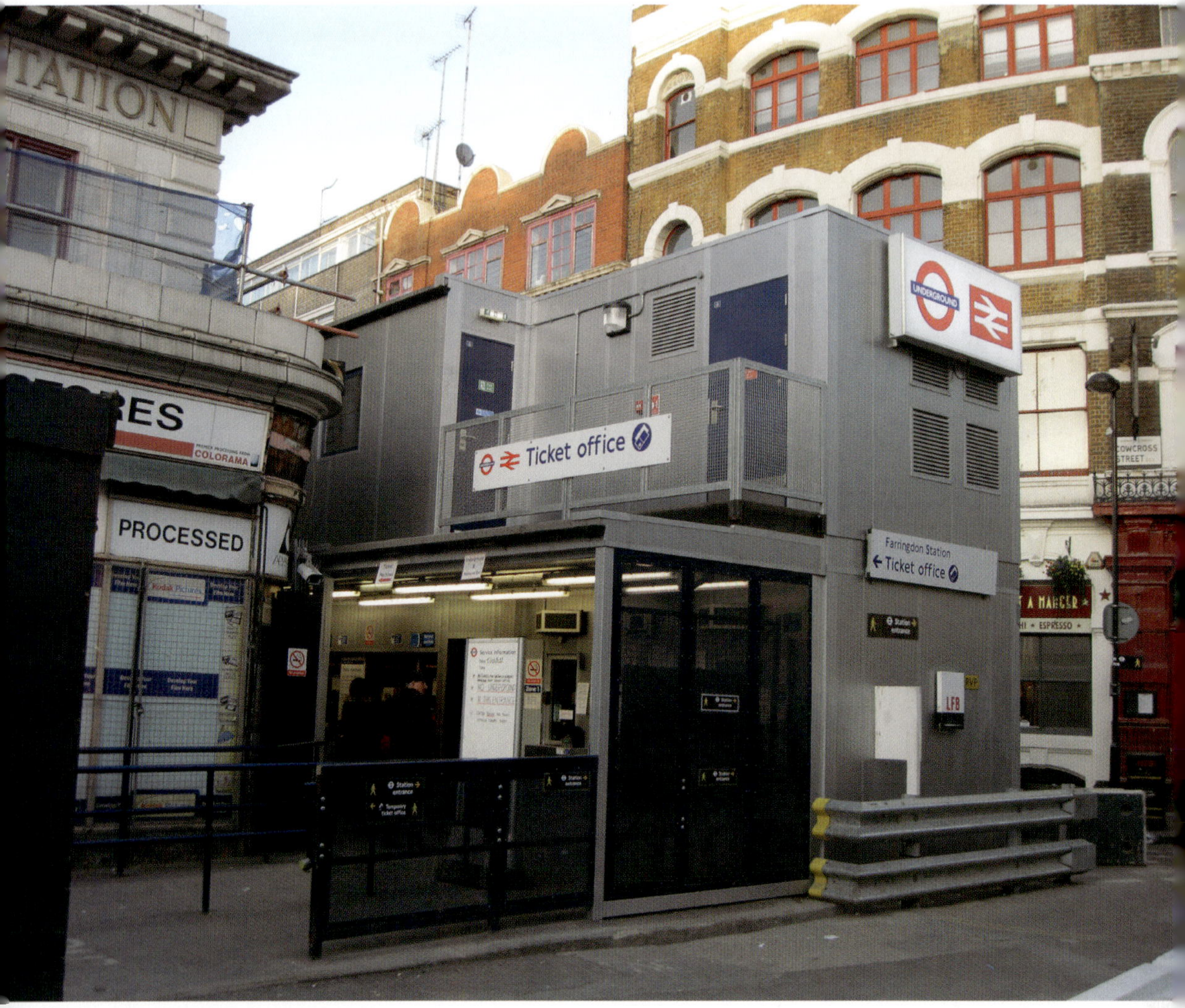

***Above*: By 12 January** 2012, rebuilding work at Farringdon Station has progressed sufficiently for a temporary booking office to have to be erected on the corner of Cowcross Street. The Metropolitan Railway buildings on the left have been fully refurbished and reopened since, but similar ones opposite them, latterly not in passenger use, were demolished and a new building opened with direct access to the extended Thameslink platforms and, later, escalators leading to the new Elizabeth Line part of the station, which opened in May 2022.

***Opposite above*: On 30 April** 2012, C77 Stock driving motor car 5720 is at the trailing end of a Barking-bound Hammersmith & City Line train calling at Barbican. The last two cars have been badly defaced by graffiti, efforts to cleanse it having also taken off some of their blue paint.

***Opposite below*: Even worse** is the damage caused to the paintwork of Metropolitan Line A60 driving motor car 5034 heading an Amersham train arriving at Barbican on the same occasion. Its paintwork was reinstated, despite these units now being rapidly withdrawn, as will be illustrated later.

5720
Barking

AMERSHAM
5034
4 1 2

***Above*: Charles Holden's** splendid ticket hall and platform buildings at Rayners Lane Station form a backdrop to this view of a Piccadilly Line 1973 Stock train terminating there also on 30 April 2012. Driving motor car 164 is leading it; a Metropolitan A Stock train from Uxbridge is at the other platform.

***Opposite above*: Representing an** earlier generation of tube stock, double-ended Acton Works depot shunter L11 was created in 1964 from Standard Stock driving motor cars 3080 and 3109. It has been rescued from Ruislip Depot's sidings and, on 24 May 2012, stands on a siding at the southern end of Epping Station. It remains on display there today and has since been nicely restored.

***Opposite below*: Rebuilding work** is still in progress at Farringdon Station, as A60 Stock driving motor car 5090 arrives heading an afternoon Metropolitan Line train to Chesham on 19 June 2012. Regular through services to that point had recently been established, replacing the earlier shuttle service to and from Chalfont & Latimer.

CHESHAM
5090
404

***Above*: With fellow** A60 driving motor car 5061 at its trailing end, the same train has taken me to Northwood, from where it departs with one of its replacements, a train of S8 Stock, in the other platform. This station is another that was substantially rebuilt as a result of the Metropolitan Line four-tracking north of Harrow on the Hill in the late 1950s/early 1960s.

***Opposite above*: An unlikely** spectacle in Northwood's leafy suburban streets on the same day is A60 Stock driving motor car 5108 being taken away for scrap. Sidings at the southern end of Northwood Station were the last stop for the Metropolitan Line's A Stock fleet, from where they were taken away by road, usually two at a time. They had been the Underground's longest-serving electric passenger stock, though that record is likely to be beaten by the Bakerloo Line's 1972 Mk2 Stock within a few years.

***Opposite below*: Whilst the** inner London suburbs of Muswell Hill and Crouch End are without rail transport, having been denied their link to the Northern City Line which was so close to completion eighty years ago, the Buckinghamshire market town of Chesham enjoys a direct link to Central London, thanks to the Metropolitan Line. This view of the terminus, showing contemporary London Underground publicity also on 19 June 2012, finds a train of A Stock awaiting departure for Aldgate.

Northwood
(B 469)
Watford
A 4125
7'-0"
Allelys
Heavy Haulage
DAF
XF
W948 RWP
STGO
CAT 3

CHESHAM
Waiting a moment
won't hurt you
Take care on the Underground
Stand back
WiFi approaching

***Above*: Having terminated** at Tower Hill on 26 July 2012, this driving motor car of District Line D Stock (7007) has been specially adorned with the Olympic symbol to celebrate the event. The Olympic torch had been carried on it a couple of days previously, between Wimbledon and Wimbledon Park Stations.

***Opposite above*: Framed by** Charles Holden's impressive Uxbridge Station, Metropolitan Line A62 Stock driving motor car 5189 has terminated there on the evening of 26 July 2012. The station was relocated eighty years previously in connection with the Piccadilly Line's extension to the town.

***Opposite below*: Also rebuilt** to one of Holden's designs at the same time, Ruislip Manor has spacious stairwells leading down from the platforms to street level, with station names unusually placed between the now disused advertisement spaces on their walls, as this evening view shows.

UXBRIDGE
5189
443

RUISLIP MANOR

***Above*: It is** almost midnight on the same day, as this train of Metropolitan Line A Stock has brought me home from the Uxbridge branch and terminated in one of Baker Street's two bay platforms. Time is running out for these trains, which still look very smart indeed after their somewhat overdue refurbishment in the late 1990s. What a change from their scruffy appearance during the Thatcher era!

***Opposite above*: Exactly two** months later, on 26 September 2012, the last train of A Stock ran in normal service on the Metropolitan Line. This view, looking down from the station footbridge, finds it at Aldgate Station awaiting departure for Amersham. Driving motor car 5062 and trailer 6062 are nearest the camera. A train of replacement S8 Stock arrives on the adjacent platform.

***Opposite below*: Complete with** commemorative headboard, A60 Stock driving motor car 5063 heads the last train in normal service when terminated at Watford Station. A new S8 unit is stabled in the siding on the left. It is about to depart for Harrow on the Hill on its very last run and will run empty back to Neasden Depot from there.

ALDGATE
5062

HARROW
5063
A STOCK
1960
2012
LAST DAY IN SERVICE

***Above*: Just three** days later, a farewell railtour to the A Stock fleet ran on 29 September 2012. A60 driving motor car 5034 heads it northbound at Baker Street Station. It looks very smart indeed, the cab having been freshly repainted, as referred to earlier.

***Opposite above*: The next** victims of new S Stock were the C69 and C77 Stock units working the Circle, Hammersmith & City and District (Edgware Road to Wimbledon section) Lines. On 30 August 2013, C77 driving motor car 5723 is bound for Hammersmith when arriving at Bromley-by-Bow Station. This station's original LT&SR booking office was replaced in 1970 as a result of the GLC's east cross route road 'improvements'.

***Opposite below*: 2013 saw** the commemoration of the 150th anniversary of London's first Underground line (the Metropolitan Railway between Paddington and Farringdon Street), and a train of new S7 Stock was adorned in this special livery. It too is working a Barking to Hammersmith service at Bromley-by-Bow on 30 August.

5723

150
150

***Above*: By now,** the two outer platforms at Whitechapel Station are used for through District and Hammersmith & City Line services. Also on 30 August 2013, C77 Stock driving motor car 5705 arrives heading one of the latter. The area that was occupied previously by the through tracks has been filled in to create a much wider passenger circulating area, necessary to facilitate interchange with the East London Line and Elizabeth Line.

***Opposite above*: On 1 September** 2013, the first train of S Stock, headed by S7 driving motor car 21317, entered service on the District Line, Here, it awaits departure from Kensington Olympia Station on its first passenger-carrying trip, which terminated at West Ham.

***Opposite below*: At West** Ham Station itself that day, driving D78 Stock driving motor car 7129 is at the trailing end of a westbound District Line service. The LT&S platform is on the left.

21317
KENSINGTON (OLYMPIA)

7129
7129
District line
051

***Above*: Another westbound** District Line train, headed by D Stock driving motor car 7028, departs from Bromley-by-Bow Station, as one of its fellows arrives heading east on 29 June 2014. By now, these were the last traditional sub-surface trains in service with London Underground.

***Opposite above*: Numerically the** first of the fleet of District Line D78 Stock driving motor cars, car 7000 is at the trailing end of the eastbound train in the previous picture. The first of these entered service early in 1980, and the last was withdrawn in the spring of 2017.

***Opposite below*: On the** same day D Stock driving motor car 7107 is at the trailing end of a District Line service departing from Barking Station for the west. Following withdrawal by London Underground, many D Stock cars have been refurbished, with various modes of propulsion, for use on various lines in Britain, most notably replacing 1938 Stock on the Isle of Wight early in 2021.

BROMLEY-BY-BOW
District line
7000
002

District line
7107
002

***Above*: Also on** 29 June 2014, a farewell tour was arranged to mark the withdrawal of the last C69/77 Stock trains, all now replaced by new S7 Stock. This view of C69 driving motor car 5578 at its trailing end sees in departing from the reversing siding east of West Ham Station.

***Opposite above*: At Barbican** Station on the same day, S7 Stock driving motor car 21367 heads a Circle Line train heading for Hammersmith, whilst an S8 Stock unit with car 22018 nearest the camera heads for Aldgate on a Metropolitan Line service. The difference between the two types, which as their notation suggests are seven- and eight-car units respectively, is that the S8s have some transverse seats and are unique to the Metropolitan Line, whilst the S7s have entirely peripheral seating. They work Circle, District and Hammersmith & City Line services. This view also shows the disused former Metropolitan Widened Lines tracks on the right. Services on them ceased in March 2009, since the lengthening of Farringdon Station's Thameslink platforms severed their connection there. However, at the time of writing, work progresses to link the tracks as far as the eastern end of Barbican Station with the existing Underground tracks at Farringdon, in order to provide stabling for S Stock units at off-peak times.

***Opposite below*: On 20 September** 2021, a new line map is displayed on the southbound platforms at East Finchley Station, showing the Northern Line's new branch from Kennington to Battersea Power Station which opened that day. The branch comes off the existing Kennington loop and can only be accessed from the Charing Cross branch.

22018
21367

1 Morden via Bank 1 min
3 Morden via Bank 7 mins
13:18:30
Northern line
Southbound platform 4
If no through train is indicated take the first train and change at Camden Town
Edgware
Burnt Oak
Colindale
Hendon Central
Brent Cross
Golders Green
Hampstead
Belsize Park
Chalk Farm
East Finchley
Highgate
Archway
Tufnell Park
Kentish Town
Camden Town
Mornington Crescent
Euston
Warren Street
Goodge Street
Tottenham Court Road
Leicester Square
Charing Cross
Embankment
Waterloo
King's Cross St. Pancras
Angel
Old Street
Moorgate
Bank
London Bridge
Borough
Elephant & Castle
Kennington
Nine Elms
Battersea Power Station
Oval
Stockwell
Clapham North
Clapham Common
Clapham South
Balham
Tooting Bec
Tooting Broadway
Colliers Wood
South Wimbledon
Morden
BE CONSIDERATE TO OTHERS
£55
allplants

***Above*: A train** of Northern Line 1995 Stock has taken me to Battersea Power Station Station on its first day and awaits return north – it will go to Mill Hill East. The two new stations are similar at platform level to those on the Jubilee Line extension, and the Elizabeth Line, but unlike them, do not have doors at their platform edges to prevent people from falling on to the track.

***Opposite above*: This huge** Underground sign outside Battersea Power Station Station's street-level building (visible on the left) leaves no doubt as to where the new station is! The power station itself is behind it on the right, alongside new residential developments.

***Opposite below*: The only** intermediate station on the new branch is Nine Elms, also pictured on its first day in use, whose Underground symbols in the windows above its entrances do not look very prominent. The there is a lot of new development here too, some of it on the site of the former LSWR Nine Elms motive power depot, which closed in 1967.

UNDERGROUND
The new Northern line extension to Nine Elms and Battersea is now open

NINE ELMS STATION
NINE ELMS STATION
NINE ELMS STATION

A very different Northern Line feature pictured that day is the up-only escalator at Highgate Station. This leads up from the top landing of the main escalators to and from the platforms (and from the former ticket office area) to Archway Road, where a small exit-only building is located. The whole affair is very much a ghost of the uncompleted Northern Line extensions, since at its exit there should have been a fully-fledged station entrance and booking office, topped with a statue of Dick Whittington and his cat. There should also have been a down escalator, as the positioning of this one makes obvious. It was not installed until 1957 and the space where the second escalator should be on the left is partly used for storage.